DNA Methylation Mechanism

Edited by Metin Budak and Mustafa Yıldız

Published in London, United Kingdom

IntechOpen

Supporting open minds since 2005

DNA Methylation Mechanism
http://dx.doi.org/10.5772/intechopen.78125
Edited by Metin Budak and Mustafa Yıldız

Part of IntechOpen Book Series: Biochemistry, Volume 15
Book Series Editor: Miroslav Blumenberg

Contributors
Metin Budak, Motoko Unoki, Aleksandra Stanković, Gaohong Zhu, Lijun Xie, Carlos M. Rodriguez Lopez, Kendall R. Corbin, Carol Bernstein, Harris Bernstein, Jorge Sandoval-Basilio, Sofia L. Alcaraz-Estrada, Nicolás Serafín-Higuera, Mónica Sierra Martínez, Claudia E. Millán-Testa, Gabriela Leija-Montoya, Silvia García, Magali Blanco-Morales, Mike J. Wilkinson, Benjamin T. Mayne, Moumouni Konate, Eileen S. Scott, Bettina Berger

First published in London, United Kingdom, 2020 by IntechOpen
IntechOpen is the global imprint of INTECHOPEN LIMITED, registered in England and Wales, registration number: 11086078, 7th floor, 10 Lower Thames Street, London, EC3R 6AF, United Kingdom
Printed in Croatia

British Library Cataloguing-in-Publication Data
A catalogue record for this book is available from the British Library

Additional hard and PDF copies can be obtained from orders@intechopen.com

DNA Methylation Mechanism
Edited by Metin Budak and Mustafa Yıldız
p. cm.
Print ISBN 978-1-83880-817-4
Online ISBN 978-1-83880-818-1
eBook (PDF) ISBN 978-1-83880-819-8
ISSN 2632-0983

For EU product safety concerns:
IN TECH d.o.o., Prolaz Marije Krucifikse Kozulić 3, 51000 Rijeka, Croatia, info@intechopen.com or visit our website at intechopen.com.

IntechOpen Book Series
Biochemistry
Volume 15

Metin Budak, MSc, PhD is an Assistant Professor at Trakya University, Faculty of Medicine. He has been Head of the Molecular Research Lab at Prof. Mirko Tos Ear and Hearing Research Center since 2018. His specializations are biophysics, epigenetics, genetics, and methylation mechanisms. He has published around 23 peer-reviewed papers, 2 book chapters, and 28 abstracts. He is a member of the Clinical Research Ethics Committee and Quantification and Consideration Committee of Medicine Faculty. His research area is the role of methylation during gene transcription, chromatin packages DNA within the cell and DNA repair, replication, recombination, and gene transcription. His research focuses on how the cell overcomes chromatin structure and methylation to allow access to the underlying DNA and enable normal cellular function.

Mustafa Yildiz is a researcher in the Department of Biophysics, Faculty of Medicine, Trakya University. He holds a BSc degree from Sakarya University (2009) and an MSc degree from Trakya University (2016). He quickly became interested in epigenetic control of gene expression which governs virtually all cellular processes such as proliferation, development, differentiation, and interaction with the environment. During his master's thesis, he studied gene expression changes in the exposure of nanoparticles and he investigated how different signaling pathways regulate genes in human cells after photodynamic therapy. He has published over 8 articles in refereed journals, 1 chapter and 32 abstracts in scientific meetings.

Editors of Volume 15:
Metin Budak and Mustafa Yildiz
Trakya University, Turkey

Book Series Editor: Miroslav Blumenberg
NYU Langone Medical Center, New York, USA

Scope of the Series

Biochemistry, the study of chemical transformations occurring within living organisms, impacts all of life sciences, from molecular crystallography and genetics, to ecology, medicine and population biology. Biochemistry studies macromolecules - proteins, nucleic acids, carbohydrates and lipids –their building blocks, structures, functions and interactions. Much of biochemistry is devoted to enzymes, proteins that catalyze chemical reactions, enzyme structures, mechanisms of action and their roles within cells. Biochemistry also studies small signaling molecules, coenzymes, inhibitors, vitamins and hormones, which play roles in the life process. Biochemical experimentation, besides coopting the methods of classical chemistry, e.g., chromatography, adopted new techniques, e.g., X-ray diffraction, electron microscopy, NMR, radioisotopes, and developed sophisticated microbial genetic tools, e.g., auxotroph mutants and their revertants, fermentation etc. More recently, biochemistry embraced the 'big data' omics systems.

Initial biochemical studies have been exclusively analytic: dissecting, purifying and examining individual components of a biological system; in exemplary words of Efraim Racker, (1913 - 1991) "Don't waste clean thinking on dirty enzymes." Today however, biochemistry is becoming more agglomerative and comprehensive, setting out to integrate and describe fully a particular biological system. The "big data" metabolomics can define the complement of small molecules, e.g., in a soil or biofilm sample; proteomics can distinguish all the proteins comprising e.g., serum; metagenomics can identify all the genes in a complex environment e.g., bovine rumen. This Biochemistry Series will address both the current research on biomolecules, and the emerging trends with great promise.

Contents

Preface XI

Section 1
Environmental Factors in Methylation Mechanisms 1

Chapter 1 3
Radiation and DNA Methylation Mechanisms
by Metin Budak

Section 2
Methylation Mechanisms in Embryonic Period 19

Chapter 2 21
Demethylation in Early Embryonic Development and Memory
by Carol Bernstein and Harris Bernstein

Chapter 3 39
Global DNA Methylation as a Potential Underlying Mechanism
of Congenital Disease Development
by Aleksandra Stanković

Chapter 4 61
Recent Insights into the Mechanisms of *De Novo* and Maintenance of
DNA Methylation in Mammals
by Motoko Unoki

Section 3
Methylation Mechanisms in Process of Cancer 75

Chapter 5 77
DNA Hydroxymethylation in the Regulation of Gene Expression
in Human Solid Cancer
*by Sofia L. Alcaraz-Estrada, Gabriela Leija-Montoya,
Nicolás Serafín-Higuera, Silvia García, Claudia E. Millán-Testa,
Mónica Sierra-Martínez, Magali Blanco-Morales and Jorge Sandoval-Basilio*

Chapter 6 97
Research Progress of DNA Methylation in Thyroid Cancer
by Zhu Gaohong and Xie Lijun

Section 4
Experimental Methods in Methylation Researchs **117**

Chapter 7 **119**
Atlas of Age- and Tissue-Specific DNA Methylation during Early Development of Barley (*Hordeum vulgare*)
by Moumouni Konate, Mike J. Wilkinson, Benjamin T. Mayne, Eileen S. Scott, Bettina Berger and Carlos M. Rodríguez López

Chapter 8 **141**
Library Preparation for Whole Genome Bisulfite Sequencing of Plant Genomes
by Kendall R. Corbin and Carlos M. Rodriguez Lopez

Preface

DNA, which is the basic determinant of life, is a molecule that has very interesting properties and can be complex and simple simultaneously. Since Watson and Crick, who discovered the molecular structure of DNA, we still continue to understand DNA.

One of these interesting features of DNA is epigenetic modifications. Non-inherited molecular changes that occur without any change in DNA sequence are called "epigenetics". Epigenetics, the existence of which was known before Watson and Crick, was first described by Conrad Waddington in 1942 as phenotypic changes caused by the relationship between gene and gene products. Although it was then thought to be related to cell division only, it has been shown that epigenetic processes can be passed on to the next generations. It is now known to have a similar functioning in yeasts, fruit flies, mice, humans, and especially plants. The genome is the sum of genetic information in the DNA sequence of a cell/organism. The epigenome is when a cell/organism regulates gene expression independently from the DNA sequence by chemical modifications of DNA and histone molecules.

The purpose of this book is to cover the mechanisms of action of epigenetic changes in different living species while also looking at the environmental factors, cancer, and embryonic life for a better understanding of epigenetic changes. It is also to create a resource book that can appeal to people from many different fields of science, from beginners to professionals, to investigate epigenetic mechanisms.

Metin Budak and Mustafa Yildiz
Trakya University,
Turkey

Section 1

Environmental Factors in Methylation Mechanisms

Chapter 1

Radiation and DNA Methylation Mechanisms

Metin Budak

Abstract

There are two types of radiation, namely ionising and non-ionising radiation, the former of which interacts with atoms or molecules while the latter does not. Technological advances, evolving with the development of mankind, have led to gradually increasing exposure to radiation. Radiation-related influences affect the cells that make up organisms in different ways, which means that they result in various effects in the affected organism. DNA methylation altered by radiation is one of the cellular systems affected in this context. DNA methylation is a major epigenetic mechanism that is particularly associated with cellular radiosensitivity, and it may also be associated with increased resistance to radiotherapy or chemotherapy. There is increasing scientific evidence that support this notion across a variety of study types from those that involve plants to those conducted with human subjects. Recent results with an increasing trend are available in this field. Our aim in this chapter is to describe the radiation environment and increasing exposure among humans as well as other living species, and to shed light on the effects of radiation on epigenetic mechanisms based on relevant scientific studies.

Keywords: ionising radiation, non-ionising radiation, epigenetic, DNA damage, methylation

1. Introduction

There are two types of radiation, namely ionising and non-ionising radiation, the former of which interacts with atoms or molecules while the latter does not. Radiation exposure is known to be gradually increasing among humans, and such exposure affects organisms in different ways. DNA methylation, a major epigenetic mechanism, is one of the cellular systems affected by radiation and of particular importance, it is involved in cancer in that it may alter sensitivity or resistance to radiotherapy at cellular level [1, 2]. In physics, radiation is defined as the emission or transmission of energy in the form of waves or particles through space or through a material medium. This is generally divided into two main types. The first type includes radiowaves, microwaves, visible light, ultraviolet and infrared. Radiation such as X-rays and gamma-radiation shows wave characteristics, while the other type of radiation refers to particles such as alpha, beta and neutron particles in electromagnetic spectrum (**Figure 1**) [3, 4]. In addition to these, there is also wave-like radiation in sound radiation and in the magnetic field of the earth. Ionising radiation is the type that has the power to displace electrons in the orbits of the atoms they encounter. Examples of this type of radiation include medical imaging devices

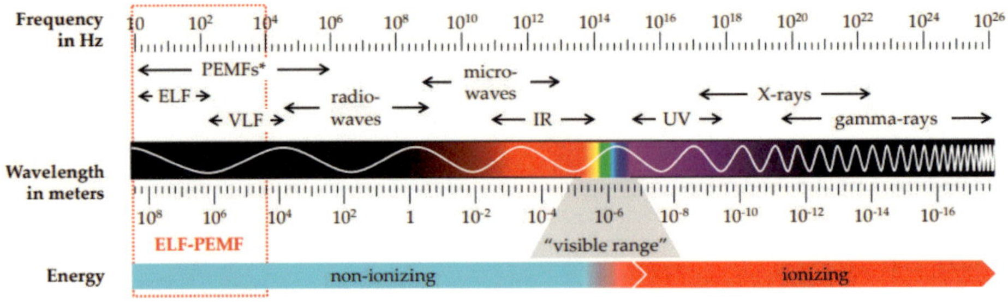

Figure 1.
Electromagnetic spectrum [6].

such as X-rays, tomography and PET (positron emission tomography) and certain medical applications including radiotherapy devices as well as the security X-rays, alpha, beta and gamma rays utilised in imaging systems used in airports and shopping malls, etc. [5]. Other types of radiation we often encounter and use include radio-TV waves, microwaves, visible light, ultraviolet and infrared radiation. All these radiation types have a variety of effects on individual species and living cells, depending on the dose and duration of exposure.

2. Effects of radiation on the cell

The effects of radiation on the cell are dose-dependent, and the radiation dose is expressed in Grey (Gy) units. Gy is briefly defined as the energy absorbed by a substance. However, spatial distribution of the irradiated volume is not taken into consideration in this definition. On the other hand, linear energy transfer (LET) is more important in radiobiology. LET is the amount of energy transferred by an ionising particle to the material traversed per unit distance, and homogeneous distribution can be achieved with low-LET, while heterogeneous distribution may occur with high LET (particulate energy). As a result of the energy transfer, different damages occur in the cell. DNA is the main target of radiation in the cell. Under normal circumstances, DNA breaks can be repaired within minutes or hours, and such breaks do not result in cell death. However, cell death may occur after double-strand breaks. The effect of radiation on DNA can be classified in two categories, i.e., direct and indirect effects [7, 8].

2.1 Direct effect

When ionising radiation strikes an atom or molecule and breaks electrons from the atom to form ionisation, direct interaction occurs. Alpha, beta and high-dose gamma rays, both with low and high unit-distance energy transfer (LET), ionise a given molecule at the point of radiation impact of radiation. This results in formation of two adjacent reactive parts in DNA structure. There may be no resultant damage if these two fragments are immediately reassembled to form the same original molecule. However, in a large macromolecule such as DNA, bond fractures may occur with this direct effect. Ionising radiation directly acting on DNA may result in open purine rings, or phosphodiester bonds may be broken, or breaks may occur in single- or double-strand DNA. On the other hand, ionising radiation is utilised in medicine, especially in nuclear medicine and radiation oncology clinics, for the treatment of cancer patients by destroying cancer cells [9, 10].

2.2 Indirect effect

Radiation can interact with molecules in the body without directly affecting DNA, such effects include ionising molecules and forming free radicals. The effect of these reactive free radicals on DNA is defined as an indirect effect. A free radical is a highly reactive atom or molecule containing one or more unpaired electrons. The unpaired single electron imparts characteristic chemical properties to the free radical, which translate into toxic effects in living cells. For example, ionised radiation can be delivered through free radicals by acting on water molecules that are present abundantly in the human body [11, 12].

The disintegration of water with radiation (radiolysis) occurs as follows: (1) H_2O + IR (ionising radiation) → e^- + H_2O^+, (2) e^- + H_2O → H_2O^-, (3) H_2O^- → OH^- + H^+, (4) H_2O^- → H^+ + OH^-. This reaction results in formation of four free radical products, i.e., H •, OH •, H^+ and OH^- (**Figure 2**) [13, 14]. Ionising particles react with DNA and cross-linking results in breakage of chemical bonds as well as structural breakdown. In the presence of oxygen, radiation produces highly destructive reactions within the cell. As a result of the indirect interaction of these toxic chemical structures with DNA, the cell may repair itself and continue to live, may fail to repair itself and die (apoptosis), or the repair may fail and lead to a mutation in the cell. With the indirect effect of ionising radiation, DNA damage can be almost two-fold higher than that caused by direct effect.

The severity of this damage depends on the radiation dose. DNA base damage is the most important type of DNA damage. Thymidine is the most radiosensitive base in this regard, followed by cytosine, adenine and guanine [15]. A 100 Rad (1 Gy) dose of low-LET radiation can produce 60–70 double-strand breaks and 1000 single-strand breaks per cell. Simple single- or double-strand breaks are responsible for cell death. Damage in DNA strands is a serious cellular phenomenon [16, 17]. However, the cell is equipped with chromosomal repair mechanisms. These repair

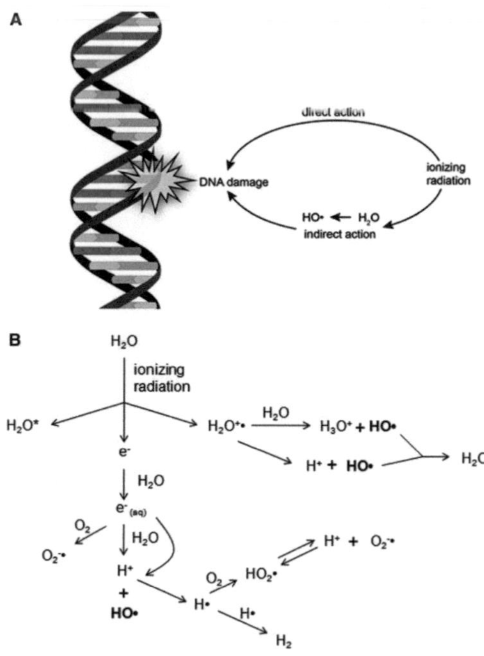

Figure 2.
(A) DNA damage by ionising of water and (B) ionise to water by ionising radiation.

mechanisms, like any other biological mechanism, are not 100% effective. DNA repairing enzymes are more effective in single-strand breaks than in double-strand breaks. If both strands of DNA are mutually damaged, they cannot repair the problem, and the damage results in cell death. Chromosomal abnormalities caused by ionising radiation generally manifest as chromosome breaks and chromatid breakage. Chromosome breaks usually occur as a result of radiation exposure in a cell during the first stage (G or early S phase) of interphase during the cell cycle [18, 19]. Chromatid breaks occur as a result of radiation exposure in the last stage of interphase (late S or G2 phase). If the chromosomal repair mechanisms fail to repair the chromosomal damage before the cell enters mitosis or meiosis, the pairing is bound to fail. This results in cell death or genetically problematic generations. Generally, cells exposed to radiation during mitosis have less time for repair; therefore, a greater number of genetic mutations and abnormal cell functions are triggered in the mitosis phase. Cells that exhibit less frequent mitotic activity (e.g., cells in the lens, nerve cells, muscle cells, skeletal cells), on the contrary, show less radiosensitivity. In radiation oncology, treatment is administered as fractions of low radiation doses (5 days, 2 Gy/day). Radiation in fractions causes both single- and double-strand breaks. Single-strand breaks are repaired between fractions within 0.3–3 h on average. Additionally, the repair capacity is higher in normal tissue compared to that in tumour tissue [20, 21]. Thus, normal tissues are protected during radiation therapy. There are several reasons for radiation to be delivered in fractions. These are called the five Rs: (1) radiosensitivity: Tissues within the organism exhibit different levels of radiosensitivity. (2) Repair: Cells have DNA repair mechanisms. In particular, single-strand breaks can be repaired rapidly by highly complex biological mechanisms. (3) Repopulation: Cells have the opportunity to replicate in between the fractions. During fractions, hypoxic cells may regain oxygen and become more susceptible to radiation. Radiation can be applied without interruption during radionuclide treatment. Both single- and double-strand breaks may occur. Within this continuity, single-strand breaks can be repaired. Radionuclide treatments may also be administered in fractions. For example; the radioactive iodine treatment, Lu-177 octreotide, and prostate-specific membrane antigen treatments are given at intervals of 6–8 weeks with intervals of 3–6 months. However, there is no scientific basis for such fractionation [22–24]. Effect of Radiation on the Cell Membrane: The main function of cell membranes is to control intracellular and extracellular substance exchange. Radiation affects the double-layer lipid structure of the cell membrane, and ionisation of membrane proteins inactivates all transport mechanisms by inactivating associated molecules. Oxidation of unsaturated molecules in their compounds with oxygen forms free radicals in double bonds and carbonyl groups, and this mechanism interacts with other organic molecules by intracellular chain reactions to convert those molecules into free radicals. There are various defence mechanisms in the body to slow down and stop this chain reaction. The Effect of Radiation Outside the Cell: There are no cells that are completely resistant to radiation. Each cell has a different level of sensitivity to radiation. While radiosensitivity is higher in frequently dividing and slightly differentiating cells (ovarian and testicular germinal cells, haematopoietic system cells, epithelial cells of the gastrointestinal system), non-dividing and highly differentiating cells (liver, kidney, muscle, nerve cells) are less sensitive to radiation [25, 26].

Effects of radiation on different types of cells: (1) change in blood parameters: Generally, a decrease in blood components may occur after a gamma dose of 500 mGy (500 rad). (2) Symptoms in the blood-producing system: Doses around 200 mGy (2 Gy) cause damage in the bone marrow, while doses above 4–6 Gy may result in complete destruction. Bone marrow may sometimes repair itself and survive at these doses; however, bone marrow repair is impossible at 7 Gy and above. (3) Symptoms in the digestive system: Doses of and above 10 Gy in whole body irradiation cause intestinal exfoliation. (4) Symptoms in the central nervous system: Doses of and above 20 Gy and in whole body irradiation cause loss of consciousness within a few hours or days. Extracellular effects of radiation are classified as deterministic effects and stochastic effects [27–30]. Deterministic Effects: Exposure to high-dose ionising radiation may result in sudden death, particularly owing to the rapid effect on bone marrow and the digestive tract. Individuals may survive acute exposure to radiation doses of up to 5 Gy. However, exposure to radiation of 50 Gy or more results in death even if medical treatment is applied. If ionising radiation exposure affects a particular region, and not the whole body, the effect of radiation exposure varies depending on the radiosensitivity of the exposed body part as well as the type and intensity of the radiation exposure. Possible consequences include skin burns and infertility with radiation exposure of the gonadal region in men (3.5–6 Gy) and in women (2.5–6 Gy), and cataract may develop due to radiation exposure of the eye (5 Gy). The deterministic effect may occur in external radiotherapy and radionuclide treatments. Stochastic Effects: These effects can be observed in a delayed manner (somatic) in non-acute (severe) radiation exposure. In particular, the effects of doses between 0.01 Sv (1 rem) and 1 Sv (100 rem) are the subject matter of extensive research. Detailed reviews have been published by the United Nations Scientific Committee on the Effects of Atomic Radiation and the United States National Academy of Sciences, The Committee on the Biological Effects of Ionising Radiation. Delayed effects of radiation can occur either by exposure to extremely high doses of irradiation at one time, or through continuous exposure to high doses of irradiation. No threshold dose can be determined for the occurrence of harmful effects. When the relation is linear and more radiation is received, the greater the likelihood of developing radiation-related harm (nonlinear model). There is no concrete data to show that low-dose radiation exposure is the cause of cancer in humans. The effects of low-dose exposure are estimated based on animal experiments and studies in subjects exposed to high-dose radiation. Possible side effects of exposure to low-dose radiation may include cancer and genetic changes [31–34].

2.3 Factors affecting the effectiveness of radiation

The health effects of exposure to ionising radiation depend on several factors. These factors are as follows: type of radiation: any type of ionising radiation may cause detrimental effects on healthy tissue. However, different radiation types at the same dose exhibit different effects. This depends on the quality factor (Q) of the radiation in question. X-rays, β-rays and positrons (Q = 1) cause the same damage in tissues, while certain heavy particles such as alpha particles, neutrons and protons cause greater damage in biological tissues than X-rays. The quality factor for alpha particles is Q = 20 [35]. Dose received: high doses cause greater health problems. Dose rate: Low-dose and time intervals of radiation exposure make biological systems resistant. While DNA and chromosomes are exposed to multiple damage in a short period of time, the repair process in response to damage takes a longer time. Single-strand breaks in DNA can usually be repaired in less than 1 h. However, double-strand breaks are more difficult to repair. Exposed body part: Although parts of extremities, such as hands and feet, are exposed to higher radiation doses, less damage occurs in these parts compared to that in other organs and tissues, such as blood. The affected individual's age: the body becomes less susceptible to the effects of radiation as cell division decreases with age. Biological differences: The tolerance against radiation varies across individuals. The studies in this area are not sufficient to determine these differences. Heat: due to the suppression of DNA repair at low temperatures, most cells are more sensitive to radiation at high temperatures. Chemical agents: Certain natural or artificial chemical agents may affect radiosensitivity, resulting in higher damage with radiation exposure. If dissolved oxygen in the tissues can increase the stability and toxicity of free radicals [36, 37].

One of the important biological effects of ionising radiation is that it can alter the epigenome, thereby leading to changes that may be transferred from one generation to the next. Such effects of radiation may occur at the somatic or reproductive cell level. These effects are often in the form of reduced global methylation of cellular DNA. Ionising radiation can damage intracellular molecules, mainly complex molecules such as proteins, lipids and RNA, leading to double-strand breaks in DNA. Therefore, such damage may cause cell cycle arrest and when this exceeds a certain level, it may even lead the cell to apoptosis or may sometimes cause abnormal cell growth. Several types of cancer cells can be completely eliminated by radiotherapy with ionising radiation; however, some others, such as stem cells, and certain types of cancer cells with survivin protein expression may exhibit resistance. Non-coding RNAs, different histone forms and chemical changes in histones as well as DNA methylations, particularly those involving cytosine and to a smaller extent, adenine, are known to be epigenetic markers. In vertebrates, especially cytosine methylation is known to affect the chromatin structure and gene expression. Epigenetic modifications such as histone modification and non-coding RNAs may be transferred through generations via cross-transitions. Several studies have shown that epigenetic changes in the first generation exposed to environmental pollutants, such as methylone, may be passed on through approximately four generations. Although there is currently no data concerning the intergenerational transmission of genome-wide methylation changes caused by ionising radiation in vertebrates, the intergenerational transmission of methylated DNAs associated with low-dose ionising radiation has been demonstrated in invertebrates [38–40].

3. Herbal effects

Flavonoids, which often have low molecular weight, are a group of secondary metabolites that may show a broad spectrum of effects in reproductive and signalling pathways such as UV-protection, protection against phytopathogens and providing signalling pathways as well as playing certain roles in different physiological pathways. The synthesis of these molecules, usually synthesised by plants, occurs via the phenylpropanoid pathway, which forms the basis of biosynthetic pathways. These molecules are synthesised by the shikimic acid pathway, leading to the formation of p-Coumaroyl-CoA through the phenylpropanoid pathway, following the formation of aromatic amino acids containing phenylalanine. This synthesis metabolism is carried out by three enzymes, namely phenylalanine ammonia lyase (PAL), cinnamate 4-hydroxylase (C4H), and 4 coumarate-CoA (4CL). Coumaroyl-CoA is also converted to naringenin from different flavonoid molecules by interacting with chalcone synthase (CHS) and chalcone isomerase (CHI). Ultraviolet-B (UV-B) radiation, a type of non-ionising radiation, generally shows a positive effect on flavonoid biosynthesis in plants. Ultraviolet-B rays usually enhance flavonoid synthesis. Ultraviolet-B rays are the only type that lead to an increase in the synthesis of flavonoids. However, apart from ultraviolet-B, ROS of solar or non-solar sources in a wider band range affect haemostasis [41, 42]. Current stress conditions may neutralise several enzymatic antioxidants, while flavonoids can act as a secondary defence system. Due to the increasing knowledge about the effects of epigenetic mechanisms on gene expression in recent years, effective mechanisms have emerged also in this field. In particular, intracellular regulation of the increase or decrease in gene expression by methylated cytosines has been shown to be also valid in plants. A number of abiotic stress factors have been found to be effective on DNA methylation dynamics. An association has been shown between cytosine methylation and ultraviolet-B rays for demethylation of the DBR2 gene and the biosynthesis of artemisinin [42, 43].

3.1 SPACE radiation

In recent years, the National Aeronautics and Space Administration (NASA) and other countries' space agencies have begun to pursue policies to develop manned-space missions and technologies between meteors or to the Moon and Mars, or on asteroids close to the Earth, in other words, they have had to initiate such policies in order to replace the diminishing resources of the Earth. As expected, the biggest problem with these tasks is the several dangerous situations astronauts may face in deep space environment, some of which may be predicted while others tend to be unpredictable. Among these dangerous situations, the major ones include protons, radiation with high-energy (H) or high-atomic number (HZE; high-atomic number (Z) and energy) and galactic cosmic radiation (GCR). HZEs, owing to their electrical charge being +2, cause damage in the cells or tissues they encounter through ionisation. This type of radiation, which the astronauts would be exposed to during deep space missions such as travelling to Mars, has been shown to cause serious cognitive disorders [44–46]. Other radiation-related effects in space missions include increased oxidative stress in the brain, neuroinflammation and other functional and structural changes, including disruption of neuronal structures and synaptic integrity [47]. There is rather a small number of studies on the mechanisms by which space radiation may cause these effects; however, molecular mechanisms likely to produce these dramatic changes in central nervous system (CNS) functions have been relatively elucidated. Biological functions of the brain are multi-layered and multi-functional, and epigenetic mechanisms—particularly DNA methylation and histone modifications—are highly important for proper functioning, which is critical for cognition. Recent developments, especially those in the field of neuroepigenetics, have shown that permanent changes in DNA methylation can significantly affect learning skills and memory. In particular, reduced activity of the DNA methyltransferase (DNMT) enzyme by 5-azadeoxycytidine (5-aza) or Zebularine, a cytidine analogue nucleoside, has been observed as well as loss or reduction of normal memory stability, reward learning or spatial learning abilities in rats that were administered RG108, a direct DNMT inhibitor [48–50]. Studies in animals exposed to a methyl-group donor diet for the manipulation of DNA methylation have shown expression changes in glutamate receptor-associated genes, and new object recognition and fearlessness [51]. Considering the effects of DNMT inhibitor agents on major methylation enzymes in the absence of toxicity and chemical stability, it has been shown that the learning and memory of certain DNA methylation enzymes can be genetically altered through DNA methylation mechanisms [51–53]. Some studies have demonstrated that memory organisation and behaviours such as dependence may be altered by decreasing the activity of certain DNMTs that add or remove methyl-group via viruses and through the changes in the expression of 10–11 translocation methylcytosine dioxygenase (TET) enzymes. Of the several epigenetic modifications, the most investigated one is the 5-methylcytosine (5mC) modification of the cytosine in DNMT. Such modifications are mostly concentrated in the promoter regions that affect the transcription of genes [48, 54, 55]. However, scientific research shows that 5mC is dynamic and may also be concentrated along the entire DNA strand or on a particular chromosome, such as the X chromosome.

DNA methylation of the DNMT enzyme group (especially DNMT1) in dividing cells is quite important for cell differentiation. In terminally differentiated neurons that make up an adult's brain, DNMT enzymes (DNMT3a and 3b in particular) are especially important since the de novo methyltransferase activity adds methyl groups to the predetermined cytosines in the DNA making up the genome. Especially when the amount of DNMT3a expression is high in mitotic neurons, it is important for the adult brain. In addition, 5mC can be oxidised by ambient oxygen or modified by TET enzymes. Of the TET enzyme group, TET3 is the most common enzyme in CNS and is known to be closely associated with learning and memory function. Likewise, the potential importance of TET1 is related to DNA methylation motifs, which may vary according to neuronal activity. Although there is less data on TET2, it is thought to be involved in developmental processes [56–59]. 5-hydroxymethylcytosine (5hmC), a highly stable, modified and oxidised form of 5mC, is found at higher levels in the brain than other organs of the body [58]. In addition, 5hmC can be actively deaminated by DNA repair mechanisms when necessary, and, despite its stability, it can be reversed to unlabelled cytosines.

3.2 Low-dose ionising radiation and oxidative stress

Ionising radiation may exhibit ionising effects directly or indirectly on the atoms of the substances it encounters. The positively charged particles are direct ionisers since they contain enough energy to disrupt the atomic structure of the substances they encounter. These charged particles are relatively large-mass and highly effective over short distances. However, since massless and wave-like radiation such as gamma travels at the light of speed and ionising radiation travels rapidly as is the case with electrons, they leave their energy in the atoms they encounter, thereby producing charged particles. As a result of this rapid action, should they encounter biological organisms, they can directly damage biomolecules such as DNA, RNA and proteins in living cells, or form highly reactive oxygen species (ROS). It has also been reported that radiation such as Laser Direct Infrared (LDIR) has considerable effects on biological substances [32, 60, 61]. Ionising radiation can also stimulate ROS production by causing nitric oxide synthase (NO) formation in the presence of biological substances such as amino acids. This NO molecule can interact with the superoxide radical (O_2-) to produce peroxynitrite (ONOO-). Peroxynitrite is a powerful oxidant radical that can interact with biomolecules such as DNA bases, proteins and lipids. Nicotinamide adenine dinucleotide phosphate (NADPH) oxidase is another important molecular source that causes intracellular ROS production. NADPH oxidase is a complex enzyme with multiple subunits located in the membrane of activated and non-phagocytic cells [62–64]. NADPH oxidase produces superoxide anions by transporting electrons from the cell membrane to extracellular molecular oxygen through cytoplasmic NADPH. Activation of NOX4 and NOX5, members of the NADPH oxidase family, by LDIR accelerator has been associated with potential DNA damage. The effect of stimuli such as LDIR on mitochondria occurs almost through the same pathway. Mitochondria are organelles that contain a group of enzymes. A series of reactions occurring in mitochondria may also lead to formation of free radical by-products due to the escape of electrons from the mitochondria. These escaping electrons contribute to the formation of superoxide at the basal level [65, 66]. High-energy radiation affects the electron flow by increasing the electron release from mitochondria, resulting in excessive superoxide production. Furthermore, ionising radiation disrupts the function of mitochondria by inhibiting the electron transport chain (ETC) enzymes from mitochondrial proteins, including aconitase. Such LDIR-mediated mitochondrial effects increase intracellular oxidative stress levels and lead to high ROS signalling. These programmed cellular changes occur in daughter cells that are formed after cell divisions following the first exposure. Genomic instability and increased changes in non-Mendelian mechanisms after LDIR-induced changes suggest that LDIR acts through epigenetic-based mechanisms. Although studies have shown cellular alterations such as RNA expression of radiation-induced DNA methylation, histone modification and nonsense transformation of gene sequences often in cancer models, the molecular and mechanical information gained across these studies are highly applicable to several biological cellular systems. Recent studies have demonstrated that LDIR exposure can change the intracellular DNA methylation profile. Using animal models, LDIR exposure has been shown to have different dose-, gender- and tissue-specific effects on reduced global methylation [67–69]. LDIR-type radiation has been shown to cause locus-specific DNA hypomethylation in the TRAPC1, FOXC1 and LINE1 (Long Interspersed Nuclear Element-1) genes in breast cancer cells. As a result of such hypomethylation, a decrease was observed in expression levels of DNA methyltransferase enzymes such as DNMT1, DNMT3A and DNMT3B as well as methylated CpG binding proteins such as MeCP2. Similarly, LDIR has been associated with the hypomethylation and activation of LINE-1, leading to increased levels of LINE-1 expression and increased genomic irregularities as a result of enhanced LINE-1 mobilisation. The effects of LDIR on reduced global DNA methylation appear to be more favourable in control groups compared to those who work in nuclear industry, thereby inherently exposed to irradiation. In relevant studies, the amount of LINE-1 methylation was higher in irradiated workers compared to controls. In these workers, reduced global methylation is observed to be significantly greater in cellular chromosomal anomalies. Thus, LDIR-mediated reduced global methylation models indicate a connection between radiation exposure and increased genomic irregularity. Although exposure to LDIR energy causes reduction in global methylation, promoter hypermethylations have been shown to be more stable compared to global hypomethylation [70–72].

4. Conclusion

After the discovery of radiation at the end of the nineteenth century, radioactivity came into use in many disciplines and in everyday life and started to be used for human benefit in some areas. It is used for the destruction of cancer cells, especially in the field of medicine, and has been increasingly used in industry, agriculture and scientific studies in recent years. The ionised radiation has effects on DNA and cells. The type of radiation varies depending on the total energy trapped in the tissues, the energy of the radiation and the tissue properties. Ionising radiation can cause different types of damage to organic tissues depending on the dose taken. Radionuclide treatments, which have been developing and diversifying rapidly in recent years, have revealed the fact that we know well the effects of radiation on tissues and cells. In addition, while the researches of the effects of this kind of radiation, especially on epigenetic mechanisms, will be important for human health. For this reason, it is important for belief that such studies are increasing gradually. This chapter summarises the recent studies, which provide compelling evidence that ionising radiation provides a mechanistic link between LDIR and epigenetic gene regulation via ROS or other mechanisms, such as low-LET. Epigenetic changes are mediated by oxidative stress. Numerous studies have demonstrated that ROS scavengers such as n-acetylcysteine and tempol prevent epigenetic DNA methylation changes induced by oxidative stress through direct or indirect mechanisms.

Conflict of interest

There is no conflict of interest.

Author details

Metin Budak[1,2]

1 Biophysics Department, Faculty of Medicine, Trakya University, Edirne, Turkey

2 Prof. Mirko Tos Ear and Hearing Research Center, Trakya University, Edirne, Turkey

*Address all correspondence to: genomicdna2@yahoo.com; metinbudak@trakya.edu.tr

References

[1] International Commission on Non-Ionizing Radiation Protection. Protection, Guidelines for limiting exposure to electric fields induced by movement of the human body in a static magnetic field and by time-varying magnetic fields below 1 Hz, Health physics. 2014;**106**(3):418-425

[2] Lomax ME, Folkes LK, O'Neill P. Biological consequences of radiation-induced DNA damage: Relevance to radiotherapy. Clinical Oncology (Royal College of Radiologists). 2013;**25**(10):578-585

[3] International Agency for Research on Cancer. Non-ionizing radiation, Part 1. Static and extremely low-frequency (ELF) electric and magnetic fields. IIARC Monographs on the Evaluation of Carcinogenic Risks to Humans, Lyon, France; 2002;**80**:1-395

[4] I.W.G.o.t.E.o.C.R.t. Humans, non-ionizing radiation, Part 1: Static and extremely low-frequency (ELF) electric and magnetic fields. IARC Monographs on the Evaluation of Carcinogenic Risks to Humans. 2002;**80**:1-20

[5] U.N.S.C.o.t.E.o.A. Radiation, Sources and Effects of Ionizing Radiation: Sources. New York, USA: United Nations Publications; 2000

[6] Ehnert S, Schroter S, Aspera-Werz RH, Eisler W, Falldorf K, Ronniger M, et al. Translational insights into extremely low frequency pulsed electromagnetic fields (ELF-PEMFs) for bone regeneration after trauma and orthopedic surgery. Journal of Clinical Medicine. 2019;**8**(12):1-22

[7] Barendsen G. Responses of Cultured Cells, Tumours, and Normal Tissues to Radiations of Different Linear Energy Transfer. Rijswijk, Netherlands: Health Research Organization TNO; 1968

[8] Rossi HH, Zaider M. Radiogenic lung cancer: The effects of low doses of low linear energy transfer (LET) radiation. Radiation and Environmental Biophysics. 1997;**36**(2):85-88

[9] Henner WD, Grunberg SM, Haseltine WA. Enzyme action at 3-termini of ionizing radiation-induced DNA strand breaks. Journal of Biological Chemistry. 1983;**258**(24):5198-5205

[10] Topyalin N, Budak M, Ozbay N, Yildiz M, Kaner T, Aydin A, et al. A comparative histopathological and immunohistochemical study of Survivin and Ki-67 proteins in glial tumours. Biotechnology & Biotechnological Equipment. 2019;**33**(1):504-509

[11] Frohlinde DS. Comparison of mechanisms for DNA strand break formation by the direct and indirect effect of radiation. Mechanisms of DNA Damage and Repair. Boston, MA, USA: Springer; 1986. pp. 19-27

[12] Michaels HB, Hunt JW. A model for radiation damage in cells by direct effect and by indirect effect: A radiation chemistry approach. Radiation Research. 1978;**74**(1):23-34

[13] Burns WG, Sims HE. Effect of radiation type in water radiolysis. Journal of the Chemical Society-Faraday Transactions I. 1981;**77**(11):2803-2813

[14] Holley AK, Miao L, Clair DKS, Clair WHS. Redox-modulated phenomena and radiation therapy: The central role of superoxide dismutases. Antioxidants & Redox Signaling. 2014;**20**(10):1567-1589

[15] Yokoya A, Shikazono N, Fujii K, Urushibara A, Akamatsu K, Watanabe R. DNA damage induced by the direct effect of radiation. Radiation Physics and Chemistry. 2008;**77**(10-12):1280-1285

[16] Singh NP. Microgels for estimation of DNA strand breaks, DNA protein crosslinks and apoptosis. Mutation Research. 2000;**455**(1-2):111-127

[17] Miller MR, Dunham JP, Amores A, Cresko WA, Johnson EA. Rapid and cost-effective polymorphism identification and genotyping using restriction site associated DNA (RAD) markers. Genome Research. 2007;**17**(2):240-248

[18] Tucker JD. Low-dose ionizing radiation and chromosome translocations: A review of the major considerations for human biological dosimetry. Mutation Research. 2008;**659**(3):211-220

[19] Maffei F, Angelini S, Forti GC, Violante FS, Lodi V, Mattioli S, et al. Spectrum of chromosomal aberrations in peripheral lymphocytes of hospital workers occupationally exposed to low doses of ionizing radiation. Mutation Research. 2004;**547**(1-2):91-99

[20] Lorimore SA, Chrystal JA, Robinson JI, Coates PJ, Wright EG. Chromosomal instability in unirradiated hemaopoietic cells induced by macrophages exposed in vivo to ionizing radiation. Cancer Research. 2008;**68**(19):8122-8126

[21] Kuhne M, Riballo E, Rief N, Rothkamm K, Jeggo PA, Lobrich M. A double-strand break repair defect in ATM-deficient cells contributes to radiosensitivity. Cancer Research. 2004;**64**(2):500-508

[22] Liu Y. Radiolabelled somatostatin analog therapy in prostate cancer: Current status and future directions. Cancer Letters. 2006;**239**(1):21-26

[23] Kabasakal L, AbuQbeitah M, Aygun A, Yeyin N, Ocak M, Demirci E, et al. Pre-therapeutic dosimetry of normal organs and tissues of (177) Lu-PSMA-617 prostate-specific membrane antigen (PSMA) inhibitor in patients with castration-resistant prostate cancer. European Journal of Nuclear Medicine and Molecular Imaging. 2015;**42**(13):1976-1983

[24] Fendler WP, Reinhardt S, Ilhan H, Delker A, Boning G, Gildehaus FJ, et al. Preliminary experience with dosimetry, response and patient reported outcome after 177Lu-PSMA-617 therapy for metastatic castration-resistant prostate cancer. Oncotarget. 2017;**8**(2):3581-3590

[25] Borovkova M, Serebriakova M, Fedorov V, Sedykh E, Vaks V, Lichutin A, et al. Investigation of terahertz radiation influence on rat glial cells. Biomedical Optics Express. 2017;**8**(1):273-280

[26] Pasternak-Mnich K, Wrobel D, Nowacka O, Pieszynski I, Bryszewska M, Kujawa J. The effect of MLS laser radiation on cell lipid membrane. Annals of Agricultural and Environmental Medicine. 2018;**25**(1):108-113

[27] Eberlein U, Peper M, Fernandez M, Lassmann M, Scherthan H. Calibration of the gamma-H2AX DNA double strand break focus assay for internal radiation exposure of blood lymphocytes. PLoS One. 2015;**10**(4):e0123174

[28] Sangsuwan T, Haghdoost S. The nucleotide pool, a target for low-dose gamma-ray-induced oxidative stress. Radiation Research. 2008;**170**(6):776-783

[29] Pohl-Rüling J, Lettner H, Hofmann W, Eckl P, Haas O, Obe G, et al. Chromosomal aberrations of blood lymphocytes induced in vitro by radon-222 daughter α-irradiation. Mutation Research, Fundamental and Molecular Mechanisms of Mutagenesis. 2000;**449**(1-2):7-19

[30] Liu S-Z. Cancer control related to stimulation of immunity by low-dose radiation. Dose-Response. 2007;**5**(1):106-108

[31] Suto Y, Akiyama M, Noda T, Hirai M. Construction of a cytogenetic dose–response curve for low-dose range gamma-irradiation in human peripheral blood lymphocytes using three-color FISH. Mutation Research, Genetic Toxicology and Environmental Mutagenesis. 2015;**794**:32-38

[32] Cardis E, Gilbert E, Carpenter L, Howe G, Kato I, Armstrong B, et al. Effects of low doses and low dose rates of external ionizing radiation: Cancer mortality among nuclear industry workers in three countries. Radiation Research. 1995;**142**(2):117-132

[33] Tang N, Bueno M, Meylan S, Incerti S, Clairand I, Villagrasa C. Simulation of early radiation-induced DNA damage on different types of cell nuclei. Radiation Protection Dosimetry. 2019;**183**(1-2):26-31

[34] Morgan WF. Non-targeted and delayed effects of exposure to ionizing radiation: I. Radiation-induced genomic instability and bystander effects in vitro. Radiation Research. 2003;**159**(5):567-580

[35] Menzel H-G. International Commission on Radiation Units and Measurements. Journal of the International Commission on Radiation Units and Measurements. 2019;**15**(1-2):1-2

[36] Mettler FA. Medical effects and risks of exposure to ionising radiation. Journal of Radiological Protection. 2012;**32**(1):N9-N13

[37] Davenas J, Stevenson I, Celette N, Cambon S, Gardette JL, Rivaton A, et al. Stability of polymers under ionising radiation: The many faces of radiation interactions with polymers. Nuclear Instruments & Methods in Physics Research Section B-Beam Interactions with Materials and Atoms. 2002;**191**(1-4):653-661

[38] Ma S, Liu X, Jiao B, Yang Y, Liu X. Low-dose radiation-induced responses: Focusing on epigenetic regulation. International Journal of Radiation Biology. 2010;**86**(7):517-528

[39] Aypar U, Morgan WF, Baulch JE. Radiation-induced genomic instability: Are epigenetic mechanisms the missing link? International Journal of Radiation Biology. 2011;**87**(2):179-191

[40] Cinemre FBS, Aydemir B, Gulyasar T, Cinemre H, Tuten A, Yolmaz N, et al. Roles of selenoprotein K and selenium status in the etiopathogenesis of preeclampsia: Their relationships with maternal lipid peroxidation, oxidative DNA damage, and glutathione peroxidase. Trace Elements and Electrolytes. 2018;**35**(3):124-130

[41] Vogt T. Phenylpropanoid biosynthesis. Molecular Plant. 2010;**3**(1):2-20

[42] Fellenberg C, Bottcher C, Vogt T. Phenylpropanoid polyamine conjugate biosynthesis in Arabidopsis thaliana flower buds. Phytochemistry. 2009;**70**(11-12):1392-1400

[43] Steward GF, Montiel JL, Azam F. Genome size distributions indicate variability and similarities among marine viral assemblages from diverse environments. Limnology and Oceanography. 2000;**45**(8):1697-1706

[44] Parihar VK, Allen B, Tran KK, Macaraeg TG, Chu EM, Kwok SF, et al. What happens to your brain on the way to Mars. Science Advances. 2015;**1**(4)

[45] Britten RA, Jewell JS, Miller VD, Davis LK, Hadley MM, Wyrobek AJ. Impaired spatial memory performance

in adult Wistar rats exposed to low (5-20 cGy) doses of 1 GeV/n (56) Fe particles. Radiation Research. 2016;**185**(3):332-337

[46] Raber J, Marzulla T, Stewart B, Kronenberg A, Turker MS. 28Silicon irradiation impairs contextual fear memory in B6D2F1 mice. Radiation Research. 2015;**183**(6):708-712

[47] Parihar VK, Pasha J, Tran KK, Craver BM, Acharya MM, Limoli CL. Persistent changes in neuronal structure and synaptic plasticity caused by proton irradiation. Brain Structure & Function. 2015;**220**(2):1161-1171

[48] Alaghband Y, Bredy TW, Wood MA. The role of active DNA demethylation and Tet enzyme function in memory formation and cocaine action. Neuroscience Letters. 2016;**625**:40-46

[49] Day JJ, Sweatt JD. DNA methylation and memory formation. Nature Neuroscience. 2010;**13**(11):1319-1323

[50] Oliveira AM. DNA methylation: A permissive mark in memory formation and maintenance. Learning & Memory. 2016;**23**(10):587-593

[51] Tomizawa H, Matsuzawa D, Ishii D, Matsuda S, Kawai K, Mashimo Y, et al. Methyl-donor deficiency in adolescence affects memory and epigenetic status in the mouse hippocampus. Genes, Brain, and Behavior. 2015;**14**(3):301-309

[52] Nikitin VP, Solntseva SV, Nikitin PV, Kozyrev SA. The role of DNA methylation in the mechanisms of memory reconsolidation and development of amnesia. Behavioural Brain Research. 2015;**279**:148-154

[53] Cattaneo A. DNA methylation and miRNAs signatures as epigenetic mechanisms bringing the memory of childhood trauma and illness. Biological Psychiatry. 2016;**79**(9):142s-143s

[54] Hore TA. Modulating epigenetic memory through vitamins and TET: Implications for regenerative medicine and cancer treatment. Epigenomics-UK. 2017;**9**(6):863-871

[55] Biergans SD, Galizia CG, Reinhard J, Claudianos C. Corrigendum: Dnmts and Tet target memory-associated genes after appetitive olfactory training in honey bees. Scientific Reports. 2016;**6**:21656

[56] Pujol Lopez Y, Kenis G, Stettinger W, Neumeier K, de Jonge S, Steinbusch HW, et al. Effects of prenatal poly I:C exposure on global histone deacetylase (HDAC) and DNA methyltransferase (DNMT) activity in the mouse brain. Molecular Biology Reports. 2016;**43**(7):711-717

[57] Antunes C, Sousa N, Pinto L, Marques CJ. TET enzymes in neurophysiology and brain function. Neuroscience and Biobehavioral Reviews. 2019;**102**:337-344

[58] Fasolino M, Welsh SA, Zhou Z. TET and 5hmC in neurodevelopment and the adult brain. DNA Modifications in the Brain: Neuroepigenetic Regulation of Gene Expression. Cambridge, Massachusetts, USA: Elsevier; 2017. pp. 61-79

[59] Budak M, Ozkan U, Yildiz M. Identification of potential methylation regions of the Smad4 mRNA and determining primer sequences for MS-PCR with the 'Methprimer' program. International Journal of Genetics and Genomics. 2019;7(3):55

[60] Baskar R, Dai J, Wenlong N, Yeo R, Yeoh KW. Biological response of cancer cells to radiation treatment. Frontiers in Molecular Biosciences. 2014;**1**:24

[61] Squillaro T, Galano G, De Rosa R, Peluso G, Galderisi U. Concise review: The effect of low-dose ionizing radiation on stem cell biology: A

contribution to radiation risk. Stem Cells. 2018;**36**(8):1146-1153

[62] Wilson A, Menon V, Khan Z, Alam A, Litovchick L, Yakovlev V. Nitric oxide-donor/PARP-inhibitor combination: A new approach for sensitization to ionizing radiation. Redox Biology. 2019;**24**:101169

[63] Park IC, Woo SH, Park MJ, Lee HC, Lee SJ, Hong YJ, et al. Ionizing radiation and nitric oxide donor sensitize Fas-induced apoptosis via up-regulation of Fas in human cervical cancer cells. Oncology Reports. 2003;**10**(3):629-633

[64] Kim S, Jeong S. Effects of temperature-dependent optical properties on the fluence rate and temperature of biological tissue during low-level laser therapy. Lasers in Medical Science. 2014;**29**(2):637-644

[65] Alexandrou AT, Li JJ. Cell cycle regulators guide mitochondrial activity in radiation-induced adaptive response. Antioxidants & Redox Signaling. 2014;**20**(9):1463-1480

[66] Jin C, Qin L, Shi Y, Candas D, Fan M, Lu CL, et al. CDK4-mediated MnSOD activation and mitochondrial homeostasis in radioadaptive protection. Free Radical Biology & Medicine. 2015;**81**:77-87

[67] Son JC, Jeong HO, Lee EK, No SG, Park D, Chung HY. Identification of specific gene expression after exposure to low dose ionizing radiation revealed through integrative analysis of cDNA microarray data and the interactome. International Journal of Radiation Research. 2019;**17**(1):15-23

[68] Goldberg Z, Schwietert CW, Lehnert B, Stern R, Nami I. Effects of low-dose ionizing radiation on gene expression in human skin biopsies. International Journal of Radiation Oncology, Biology, Physics. 2004;**58**(2):567-574

[69] Sokolov M, Neumann R. Global gene expression alterations as a crucial constituent of human cell response to low doses of ionizing radiation exposure. International Journal of Molecular Sciences. 2015;**17**(1):1-18

[70] Erichsen L, Beermann A, Arauzo-Bravo MJ, Hassan M, Dkhil MA, Al-Quraishy S, et al. Genome-wide hypomethylation of LINE-1 and Alu retroelements in cell-free DNA of blood is an epigenetic biomarker of human aging. Saudi Journal of Biological Sciences. 2018;**25**(6):1220-1226

[71] Kloypan C, Srisa-art M, Mutirangura A, Boonla C. LINE-1 hypomethylation induced by reactive oxygen species is mediated via depletion of S-adenosylmethionine. Cell Biochemistry and Function. 2015;**33**(6):375-385

[72] Chen B, Dai Q, Zhang Q, Yan P, Wang A, Qu L, et al. The relationship among occupational irradiation, DNA methylation status, and oxidative damage in interventional physicians. Medicine (Baltimore). 2019;**98**(39):e17373

Section 2

Methylation Mechanisms in Embryonic Period

Chapter 2

Demethylation in Early Embryonic Development and Memory

Carol Bernstein and Harris Bernstein

Abstract

DNA repair processes arose early in evolution. During evolution, DNA base excision repair apparently acquired additional roles in demethylation of cytosines in DNA. Demethylation is central to two mammalian fundamental processes. Embryonic reprogramming and neuronal memory require rapid gene expression alterations depending in part on demethylations. The active demethylation reactions in both processes primarily depend, first, on the family of 5-methylcytosine oxidases sharing the acronym ten-eleven translocation (TET methylcytosine dioxygenases) and, second, on DNA base excision repair enzymes. In mice, within 6 h of fertilization, the paternal chromosomes are close to 100% actively demethylated through TET and repair activity. (Methylation of maternal DNA is blocked during subsequent cycles of replication, so methyl groups on maternal DNA, passively, becomes highly diluted over the next 4 days.) Rats subjected to one instance of contextual fear conditioning create an especially strong long-term memory. At 24 h after training, 9.2% of the genes in the rat genomes of hippocampus neurons are differentially methylated, including over 500 genes with demethylation. The emergence of embryonic development in evolution depended on preexisting DNA methylation/demethylation pathways to modify gene expression. The further emergence of memory likely evolved from the earlier set of methylation/demethylation capabilities associated with embryonic development.

Keywords: TET enzymes, OGG1, epigenetic, base excision repair, DNA repair, neuroepigenetics, neurogenesis, brain evolution

1. Introduction

DNA repair processes have a central role in epigenetic demethylation reactions that are employed in both early embrylonic development and in memory. DNA likely emerged as the genetic material as long as 3.5 billion years ago [1]. From its inception as the genetic material, DNA was likely subject to damage. In present day organisms damage to DNA is frequent and occurs due to both metabolic and hydrolytic processes [2] as well as a result of environmental agents such as UV light and ionizing radiation. Thus, enzymes promoting DNA repair likely have been retained based on their adaptive benefit since early evolution. Currently, in humans, about 169 different DNA repair proteins have been identified [3]. During the course of evolution, many of these DNA repair proteins developed more than one enzymatic capability. For instance, at least 17 DNA repair proteins act in both a DNA repair pathway and in an apoptosis pathway [4]. These dual role proteins are required for DNA repair when DNA damages are at relatively low levels but are active and required for apoptosis when DNA damages are at high levels.

In addition to the multiple roles of some DNA repair proteins, some endogenously produced DNA damaging agents also appear to have multiple roles. Reactive oxygen species (ROS) are produced by mitochondria during oxidative metabolism, and a small proportion are released from the mitochondria and interact with proteins, lipids and DNA to alter their structures. ROS can damage DNA in ways that are mutagenic or disruptive to expression. Thus, excessive ROS can cause mutations and other alterations leading to cancer [5]. However, ROS can interact with DNA to serve important positive roles. A large body of literature has shown the necessary roles of appropriate levels of ROS in embryonic development [6, 7] and in learning and memory [8, 9].

2. Demethylation in embryogenesis

During early embryogenesis of mammals, pathways of rapid demethylation are employed at multiple DNA sites to form totipotent cells. Subsequently, locally deposited methylations enable formation of subsets of cells that became specialized tissue types, such as primordial germ cells and neuronal stem cells [10]. Such rapid demethylations and subsequent methylations have also now been found to occur in the formation of memories and learning [11] and in both cases the mechanism of methyl group removal occurs by similar pathways involving TET enzymes and base excision DNA repair.

In embryogenesis, rapid and large scale demethylations occur at two stages [12]. One extensive set of demethylations occurs within a few hours after the sperm enters the egg, forming the zygote. Almost all methyl groups are removed from the paternal-origin chromosomes within 6 h of forming the zygote, before any replication has occurred [13]. Another extensive demethylation occurs early in embryogenesis, in the nuclei of the primordial germ cells shortly after they devolve from the other cells which are forming somatic tissues [14]. This stage of demethylation occurs in two phases. There is a first phase of rapid proliferation without methylation, causing dilution of methylation with a loss of methylation at almost all genomic sequences. Then there is a second phase, involving specific sites including germ-line and meiosis specific genes, where the demethylation is active and proceeds by pathways involving TET enzymes and base excision DNA repair.

Methylation of sites (which can be demethylated) in mammalian DNA are usually restricted to cytosines, forming 5-methylcytosine (5mC) (**Figure 1**). In this figure, the addition of a methyl group at the 5 position of cytosine is shown within a red oval. Of all the cytosines in DNA, the 5mCs occur primarily at "CpG" sites [16]. A CpG site is where a cytosine in a DNA strand is followed by a guanine nucleotide in the linear sequence of bases along the 5′ to 3′ direction. There are 28 million CpG sites in the human genome [17]. In humans, about 60% of the 28 million CpG sites are methylated in most somatic tissues [18]. CG dinucleotides (CpG sites) represent about 1% of total bases in the mammalian genome [19]. Three DNA methyltransferases in humans can methylate a base in DNA. These enzymes show a strong preference for methylating cytosines in CpG sites [20].

Mouse DNA is very similar to human DNA, with about 99% of mouse genes having a homolog in the human genome, and mice and humans having about the same number of genes [21]. However, the mouse sequence is about 14% shorter than the human sequence [21]. The mature mouse sperm genome has 80–90% overall methylation of its CpG sites, the highest global DNA methylation level of any cell in the mouse [12]. Because of its shorter sequence, we can speculate that there may be fewer than 28 million CpG sites in the mouse genome, perhaps 86% as many as in the human genome, or about 24 million CpG sites. Thus, of the likely 24 million CpG sites, there are about 19–22 million methylated sites in mouse sperm DNA. In mouse zygotes, partial demethylation of the paternal nucleus is already evident 3 h after formation of the zygote [13]. By 6 h, demethylation of the paternal nucleus appears to be complete (**Figure 2**). During the subsequent first mitosis, there is just a small but significant residual methylation signal in some but not all of the paternally derived chromosomes [13]. By 3–4 days after fertilization, after replication to generate 16 cells, the embryo has formed a morula (a round body of cells with no differentiation) (**Figure 2**). By this time both the paternal and maternal chromosomes have mixed together in a single nuclear area and all have very low levels of methylation (In **Figure 2**, the methylation levels of the paternal and maternal chromosome are approximately represented by the blue lines during the period they can be distinguished. When the chromosomes become mixed, after two mitoses, the methylation level of the mixed chromosomes is represented by a brown line).

The almost compete demethylation of the zygote DNA in the paternal chromosomes at 22–25 million CpG sites occurs before any DNA replication. Thus, it occurs by an active process not connected to replication. The demethylation of the maternal chromosomes appears to largely take place by blockage of the methylating enzymes from acting on maternal-origin DNA and dilution of the methylated

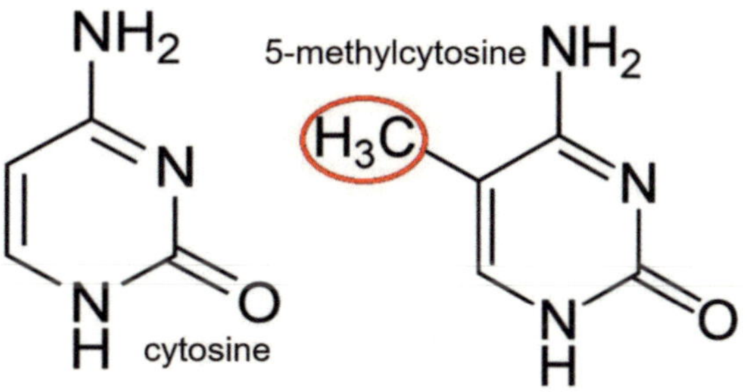

Figure 1.
DNA methylation most often is the addition of a methyl group to cytosine in DNA. The image shows cytosine and 5-methylcytosine. In mammals, DNA methylation most frequently occurs at a cytosine followed by guanine in the DNA [15].

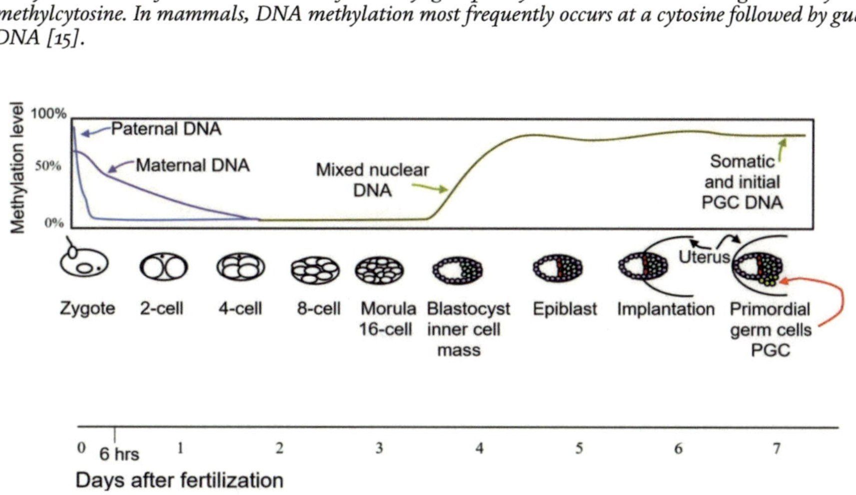

Figure 2.
Methylation levels during mouse early embryonic development.

maternal DNA during replication. At the second metaphase after fertilization, maternal chromosomes showed methylation on only one of the two sister chromatids. This sister chromatid differentiation is consistent with mostly replication-dependent passive maternal chromosome demethylation [22]. Consequently, four-cell embryos have a much lower methylation density over the maternal nuclear compartment. Methylation of the maternal genome further decreases with every additional replication cycle. The morula (at the 16 cell stage), overall, has much reduced methylation of DNA.

High levels of *de novo* DNA methylation then occur in the cells of the inner cell mass of the blastocyst, to establish the specific methylation patterns of principal cell lineages in the early embryo [13]. Afterwards, by day 5 of mouse embryogenesis, the epiblast is formed, followed by implantation of the epiblast in the uterine epithelium (**Figure 2**). By day seven after fertilization, the newly formed primordial germ cells (PGC) in the implanted embryo devolve from the remaining somatic cells. At this point the PGCs have high levels of methylation. These cells migrate from the epiblast along the hindgut toward the genital ridges starting about day 7.8. By day 8.5 they are rapidly proliferating and beginning demethylation in two waves. In the first wave, demethylation is by replicative dilution, but in the second wave demethylation is by an active process. The second wave, during days 9.5–13.5, leads to demethylation of specific loci. At day 13.5, the PGC genomes display the lowest levels of DNA methylation of any cells of the mouse in the entire life cycle [14].

2.1 Mechanisms of demethylation

The demethylation of methylated CpG sites of DNA occurs in three stages: (1) recruitment of a TET enzyme to initiate demethylation (although there is one minor mechanism that does not utilize a TET enzyme); (2) intermediate steps of oxidation or oxidative deamination (forming intermediate products of demethylation); and (3) culminating steps of DNA base excision repair resulting in final replacement of 5-methylcytosine with cytosine.

The pathways by which demethylation can occur [23] are shown in outline in **Figure 3**. This figure indicates two types of oxidation reactions that may occur in demethylation. One occurs by oxidation of the added methyl group at the 5 position of cytosine. The other occurs through oxidative deamination of the amine group at the 4 position of cytosine. The pathway on the left depends on oxidation of each of the adducts on the 5 position of cytosine, sequentially, by a TET enzyme, followed by action of base excision repair (BER) enzymes. TET enzymes (ten-eleven translocation methylcytosine dioxygenases) oxidize adducts on cytosine in an iron and alpha-ketoglutarate dependent process. This TET-type dependent pathway likely carries out the bulk of the demethylations discussed here. However, as reviewed [25], two other pathways involving AID/APOBEC and base excision repair enzymes can occur. In one pathway there is an initial TET reaction. The other pathway involving AID/APOBEC results in oxidative deamination of 5mC directly to thymine followed by base excision repair. The activity of AID/APOBEC appears to cooperate with a TET enzyme in neuronal functions [26]. It is notable that demethylation, in all its pathways, employs the enzymes of the base excision repair pathway.

3. Base excision repair

In **Figure 3**, base excision repair is indicated by the highlighted acronym "BER". To complete the description of the mechanism shown above, we include a

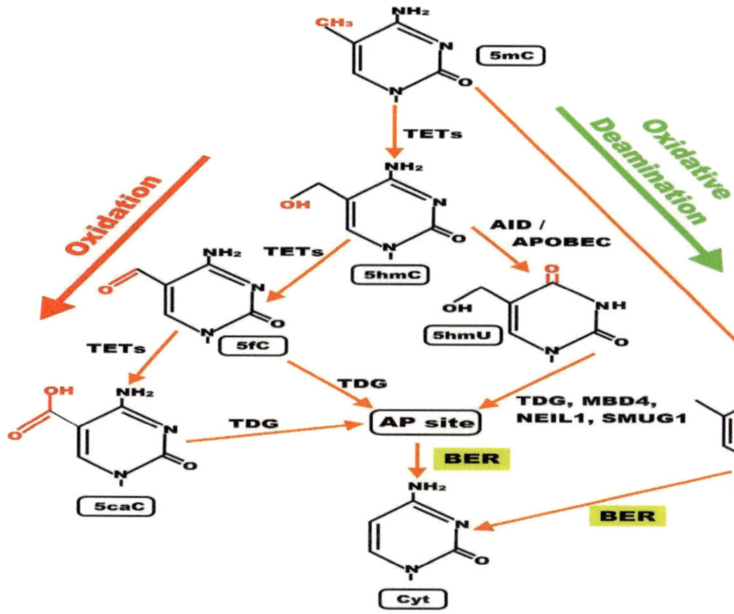

Figure 3.
Demethylation of 5-Methylcytosine (5mC) in neuron DNA. As reviewed in [23], in brain neurons 5mC is oxidized by the ten-eleven translocation (TET) family of dioxygenases (TET1, TET2, TET3) to generate 5-hydroxymethylcytosine (5hmC). In successive steps TET enzymes further hydroxylate 5hmC to generate 5-formylcytosine (5fC) and 5-carboxylcytosine (5caC). Thymine-DNA glycosylase (TDG) recognizes the intermediate bases 5fC and 5caC and excises the glycosidic bond resulting in an apyrimidinic site (AP site). In an alternative oxidative deamination pathway, 5hmC can be oxidatively deaminated by activity-induced cytidine deaminase/apolipoprotein B mRNA editing complex (AIP/APOBEC) deaminases to form 5-hydroxymethyluracil (5hmU) or 5mC can be converted to thymine (Thy). 5hmU can be cleaved by TDG, single-strand-selective monofunctional uracil-DNA glycosylase 1 (SMUG1), Nei-Like DNA Glycosylase 1 (NEIL1), or methyl-CpG binding protein 4 (MBD4). AP sites and T:G mismatches are then repaired by base excision repair (BER) enzymes to yield cytosine (Cyt). Figure by [24].

diagram illustrating the base excision repair pathway used in the latter stages of the conversion of 5mC to C (**Figure 4**). In this diagram the two strands of DNA are represented by parallel horizontal lines. With the first downward arrow we show thymine DNA glycosylase (TDG) removing 5-formylcytosine (5fC) from the DNA backbone, leaving an *apyrimidinic site.* Then AP endonuclease cleaves the 5′ deoxyribose-phosphate in the DNA backbone of a single strand, leaving a 3′ hydroxy end and a 5′ deoxyribose phosphate end (second downward arrow). This is followed by either short patch or long patch repair. In short patch repair, 5′ dRP lyase trims the 5′ dRP end to form a phosphorylated 5′ end. This is followed by DNA polymerase β adding a single cytosine to pair with the pre-existing guanine in the complementary strand and then DNA ligase to seal the cut strand. In long patch repair, DNA synthesis is thought to be mediated by polymerase δ and polymerase ε performing displacing synthesis to form a flap. Pol β can also perform long-patch displacing synthesis. Long-patch synthesis typically inserts 2–10 new nucleotides. Then flap endonuclease removes the flap, and this is followed by DNA ligase to seal the strand.

In an example below (see "Targeting TET to 5-methylcytosine") we show that, in at least one well documented case, the ROS-induced damage of 8-OHdG at a CpG site initiates demethylation. In the base excision pathways shown in **Figure 4**, it is not clear at what stage 8-OHdG itself may be removed. Thus, 8-OHdG is allowed to remain in most steps of this diagram.

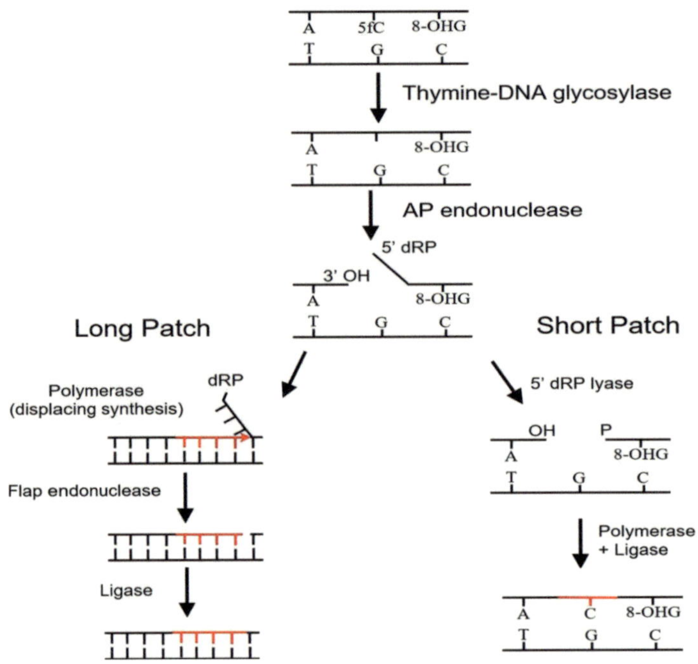

Figure 4.
An example of base excision repair of 5-formylcytosine (5fC) (adjacent to 8-OHG, an oxidized guanine) by short patch repair or long patch repair.

4. TET enzymes

As described by Jin et al. [27] and Melamed et al. [28], there are a number of TET enzymes, including at least two isoforms of TET1, one of TET2 and three isoforms of TET3. As reviewed [28], the full-length canonical TET1 isoform appears virtually restricted to early embryos, embryonic stem cells and PGCs. The dominant TET1 isoform in most somatic tissues, at least in the mouse, arises from alternative promoter usage which gives rise to a short transcript and a truncated protein designated TET1s. The isoforms of TET3 are the full length form TET3FL, a short form splice variant TET3s, and a form that occurs in oocytes and neurons designated TET3o. TET3o is created by alternative promoter use and contains an additional first N-terminal exon coding for 11 amino acids. TET3o only occurs in oocytes and neurons and was not expressed in embryonic stem cells or in any other cell type or adult mouse tissue tested [27]. Whereas TET1 expression can barely be detected in oocytes and zygotes, and TET2 is only moderately expressed, the TET3 variant TET3o shows extremely high levels of expression in oocytes and zygotes, but is nearly absent at the 2-cell stage [29].

The TET enzymes generally do not specifically bind to 5-methylcytosine except under particular conditions, such as the two conditions described below, in "Targeting TET1 to 5-methylcytosine" and in "TET in learning and memory." Without targeting, TET1 predominantly binds to high CG promoters and CpG islands (CGIs) genome-wide by its CXXC domain that can recognize un-methylated CGIs [30]. TET2 does not have an affinity for 5-methylcytosine in DNA [31]. The CXXC domain of the full-length TET3, which is the predominant form expressed in neurons, binds most strongly to CpGs modified by 5-carboxycytosine (5caC) (**Figure 3**), although it does also bind to un-methylated CpGs [28].

4.1 Targeting TET to 5-methylcytosine

One mode of recruitment of a TET enzyme to 5-methylcytosine in DNA, in order to initiate demethylation, was investigated by Zhou et al. [32]. In this mode, recruitment was found to depend on ROS treatment of cells. This finding is significant because appropriate levels of ROS are known to be needed in both embryogenesis [6, 7] and in learning and memory [8, 9]. ROS cause oxidative damages to DNA, but these damages are not random. Because electron "hole" pausing at the sites of the lowest ionization potential increases the probability of stable adduct formation, DNA oxidation tends to be sequence dependent [19]. As reviewed by Ming et al. [19], cytosine methylation increases the reactivity of guanine bases in 5mCpG dinucleotides toward electrophiles and oxidants. This is likely due to the transmission of an electronic effect from the 5mC to its partner guanine through hydrogen bonding within the 5mC:G base pair. Ming et al. [19] experimentally showed that oxidation of guanines was enhanced within endogenously methylated 5mCpG dinucleotides.

There are many types of oxidative DNA damage, but the most common endogenous oxidative damage in DNA is 8-OHdG [33]. The molecular structure of 8-OHdG is shown as part of **Figure 5**. In **Figure 5**, the structure labeled in red as "8-OHdG" is a guanine with the oxidative damage, an added OH group at the 8 position of the pentane (5-sided) ring, shown in red. 8-OHdG can be experimentally increased in cells by treatment with Hoechst dye followed by micro-irradiation with 405 nm light [34]. The irradiation can be performed along a narrow line.

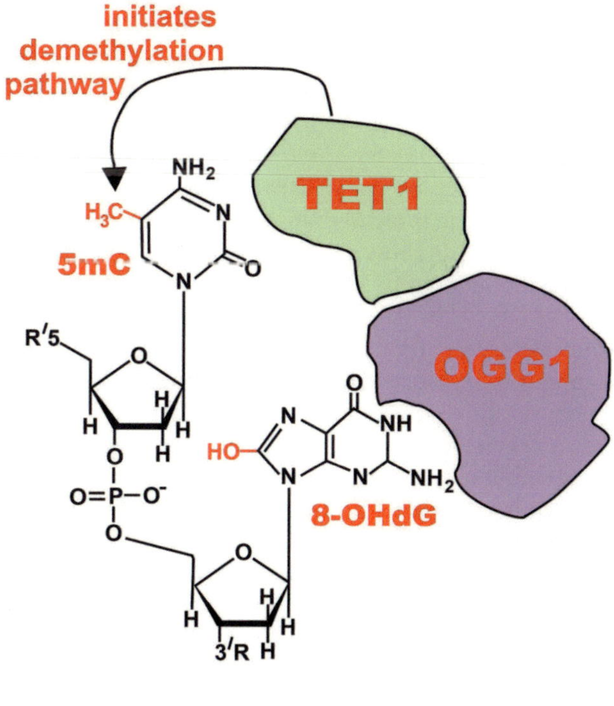

Figure 5.
Initiation of DNA demethylation at a CpG site. In adult somatic cells DNA methylation typically occurs in the context of CpG dinucleotides (CpG sites), forming 5-methylcytosine-pG, or 5mCpG. Reactive oxygen species (ROS) may attack guanine at the dinucleotide site, forming 8-hydroxy-2′-deoxyguanosine (8-OHdG), and resulting in a 5mCp-8-OHdG dinucleotide site. The base excision repair enzyme OGG1 targets 8-OHdG and binds to the lesion without immediate excision. OGG1, present at a 5mCp-8-OHdG site recruits TET1 and TET1 oxidizes the 5mC adjacent to the 8-OHdG. This initiates demethylation of 5mC [37].

Within 6 s of the irradiation with 405 nm light, there is half-maximum recruitment of OGG1 to the irradiated line. OGG1 (8-oxoguanine DNA glycosylase) is an enzyme that removes the oxidative damage 8-OHdG from DNA [35]. Removal of 8-OHdG, during base excision repair, occurs with a half-life of 11 min [36]. Thus, OGG1 protein rapidly complexes with 8-OHdG (6 s) but the OGG1-8-OHdG complex has a relatively long half-life (11 min).

H_2O_2 is a reactive oxygen species. Zhou et al. [32] treated cells in culture with 500 μM H_2O_2 for 6 h and this caused a more than 3-fold increase in 8-OHdG. The cells treated with H_2O_2 also became substantially demethylated, with methylation reduced to less than 1/4th the original methylation level. They then used cells in which OGG1 was inhibited, either by applying siRNA or by using OGG1 mutant knockout cells. In cells with inhibited or absent OGG1, treatment with H_2O_2 did not cause demethylation. These first experiments indicate that OGG1 has a role in H_2O_2 -induced demethylation.

Zhou et al. [32] examined the interaction between OGG1 and the TET enzymes that are involved in demethylation [23]. OGG1 did not interact with TET2 or TET3. However, OGG1 interacted with TET1. They found that the two proteins co-immunoprecipitated, and this co-immunoprecipitation did not depend on interactions with DNA or with 8-OHdG. Thus, OGG1 can attract or "recruit" TET1. They then used a double-stranded oligonucleotide containing 8-OHdG in solution in a pull-down assay using streptavidin beads. They found that OGG1 added to the assay could be pulled down by oligonucleotides containing 8-OHdG. TET1 could not be pulled down by oligonucleotides containing 8-OHdG, but TET1 could be pulled down if in the presence of OGG1. Their results imply that OGG1 attaches to 8-OHdG and then recruits TET1 to 8-OHdG lesions. They indicated that this could allow TET1 to initiate DNA demethylation of methylated CpGs after 8-OHdG lesions are formed (**Figure 5**). As shown in this figure, TET1 first interacts with OGG1 and then is close enough to the methyl group CH_3 (shown in red) on the 5 position of the cytosine, to initiate the oxidation of the methyl group. This mechanism is notable for likely using two co-opted elements of DNA base excision repair (BER). First, OGG1 is an initiating enzyme in BER of 8-OHdG, but acts here to recruit TET1. Second, once the intermediate products of demethylation are formed by TET1, such as 5fC or 5caC as shown in **Figure 3**, then thymine DNA glycosylase (TDG) can initiate BER as shown in **Figure 4**, and complete the demethylation of 5mC to C.

OGG1 knockout mice seem to undergo a fairly normal embryogenesis, and the young new mice appear to be mostly normal [38], though they have a deficit in learning and memory as shown by a passive avoidance test [39] and a deficiency in immune responses (reviewed in [40]). TET1 knockout mice are also viable and fertile, with no discernible morphological or growth abnormality. However, TET1 knockout mice have an impairment in spatial learning and short-term memory [41] as well as deficiencies in fear memory extinction and spacial memory extinction [42]. On the other hand, over-expression of TET1 impairs hippocampus-dependent long-term associative memory [43]. A TET3 homozygous mutation, unlike a TET1 knockout, leads to neonatal lethality [44]. Thus TET3 is essential in embryogenesis. As pointed out above, TET3 (but not TET1 and TET2) is highly expressed in oocytes and zygotes (also shown in [45]).

5. Demethylation in neurogenesis

Neurogenesis in mouse takes place starting about day 10.5 after fertilization of the egg. Early in neurogenesis, some embryonic stem cells (ESCs) begin differentiating into neural stem cells (NSCs) and neural progenitor cells (NPCs) [46]. At this point, 8% of CpGs unmethylated in ESCs become largely methylated in NPCs, whereas approximately 2% of CpGs methylated in ESCs become unmethylated [46]. These data suggest that 5mC undergoes significant dynamic changes during ESC differentiation into NSCs. As shown by Pilz et al. [47], NPCs generate neurons throughout life in the dentate gyrus of the hippocampus of mice. Zhang et al. [41] examined adult NPCs purified from wild type and TET1 knockout mice. They found that 478 genes showed elevated promoter methylation levels in TET1-null NPCs compared to the wild-type control, while only 32 genes had lower methylation. Thus, TET1 appears to function in demethylation during neurogenesis in the adult brain.

6. Demethylation in learning and memory

Learning and memory have levels of permanence, differing from other mental processes such as thought, language, and consciousness, which are temporary in nature. Learning and memory can be either slowly accumulated (multiplication tables) or rapidly (touching a hot stove), but once attained, can be recalled into conscious use for a long time. As pointed out by Alberini [48], humans can generally recall a painful fact or trauma in detail for a lifetime. Similarly, humans remember a very happy day for a long time afterwards. At least two early proposals were presented, indicating, on theoretical grounds, that the methylation and demethylation of DNA in neurons is the physical basis of memories. In 1969 Griffith and Mahler [49] published an article that made a number of salient points. They noted that, at least in man, memories may survive for periods of almost the entire lifetime. Further, DNA is the one molecule which, apart from possible minor effects due to genetic damage and repair, is surely present in neurons for the whole of the lifetime of the organism. This led them to the suggestion that the physical basis of memory could lie in the enzymatic modification of the DNA of nerve cells. They further indicated that a plausible suggestion would be that the modification consists of methylation (or demethylation) of DNA.

In 1999 Holliday [50] noted that long-term human memory can be retained for many decades. The exceptional stability required suggests that essential memory components may be based on chemical changes. He proposed that the enzymatic modification of cytosine in DNA to 5-methylcytosine may provide this necessary stability. The general model proposed is that specific sites in the DNA of neurons required for memory can exist in alternative methylated or non-methylated states. The initial signal, which is to be memorized, switches the DNA from a modified to an unmodified state, or vice versa. It should be noted that the presence or absence of DNA methylation at a particular sequence of DNA can be thought of as a 0, 1 binary code. Thus, 10 such sites have 2^{10} (1024) epigenotypes and potential phenotypes, and 30 such sites could have up to 2^{30}, or 1.07×10^9 epigenotypes. Clearly, such a set of control mechanisms has enormous potential for neuronal specificity.

One form of long-term memory, associative learning, is contextual fear conditioning [51]. As an example of contextual fear conditioning, a rodent is placed in a novel environment (a new context) and is then subjected to an electric shock (e.g. a footshock). The rodent then experiences robust fear learning, shown by a strong fear response, when the rodent is placed in that context again. Contextual fear conditioning occurs very rapidly (it can occur with a single event) and it has a lasting effect [51]. Kim and Jung [51] reviewed the evidence that the hippocampus region of the brain is where contextual fear memories are first stored, and that this storage is transient and does not remain in the hippocampus (**Figure 6**). (Note that while this diagram shows a single hippocampus in a human brain, humans have two hippocampi, one in each hemisphere of the brain.) They point out, in rats, that contextual fear conditioning is abolished when the hippocampus is subjected to hippocampectomy just 1 day after conditioning. However, the rats retain a considerable amount of contextual fear when a long delay of 28 days is imposed between the time of conditioning and the time of hippocampectomy. Using localized lidocaine injections to impede brain functions, Frankland et al. [53] showed that much of the long term storage of contextual fear conditioning memory appears to take place in the anterior cingulate cortex (**Figure 6**) (Note that there is a single anterior cingulate cortex of the human brain and it *resides in the medial wall of the two cerebral hemispheres*).

When methods to detect DNA methylation at specific locations on chromosomes became available, early experiments focused on particular genes known to be important for memory. One such gene is *PP2B* (*protein phosphatase 2B*), also known as *calcineurin* (*CaN*). This gene is of particular interest because it is the only Ca^{++}-activated protein phosphatase in the brain and a major regulator of key proteins essential for synaptic transmission and neuronal excitability [54]. Miller et al. [55] found that persistent, specific hypermethylation of the *CaN* gene in the anterior cingulate cortex was induced in rats by a single contextual fear conditioning event at a time when a long-term memory was formed. Demethylation at a specific locus also has been investigated. Brain-derived neurotrophic factor (BDNF) is known to be important in memory [56]. As reviewed by Lubin et al. [57], the *bdnf* gene consists of eight 5′ exons each linked to individual promoter regions, and a 3′ exon (IX). Lubin et al. [57] subjected rats to contextual fear conditioning. Their sequencing data confirmed active demethylation of *bdnf* exon IV after fear conditioning along with a strong increase in expression of exon IV in the hippocampus at 2 h after fear conditioning. As noted above [51], the hippocampus region of the brain is where contextual fear memories are first stored, but this storage is transient. In the experiments of Lubin et al. [57] the RNA expression of exon IV of the *bdnf* gene returned to baseline level by 24 h after the fear conditioning.

More recently, methods became available to identify differentially methylated genes in entire genomes. In 2016, Halder et al. [58] used mice subjected to contextual fear conditioning and evaluated whole neuron genomes for differentially methylated genes and for differentially expressed genes. In one part of their study they looked at

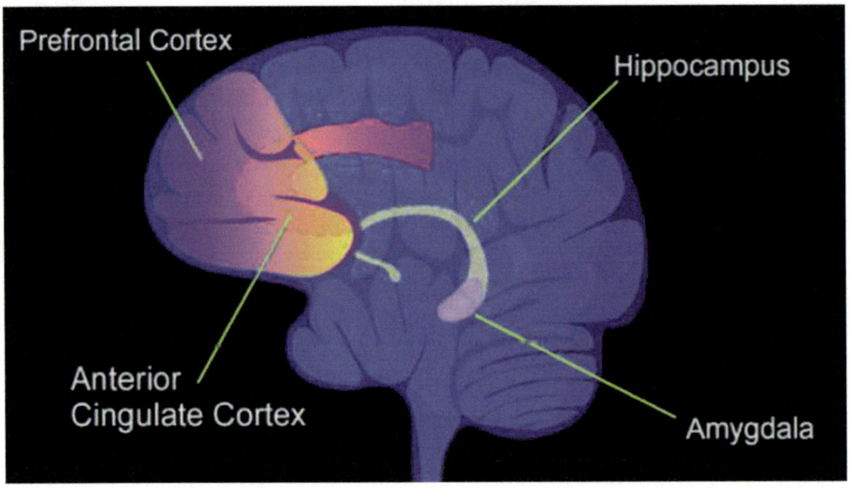

Figure 6.
Some regions of the brain involved in memory [52].

the hippocampal CA1 region, a region that is crucial for short-term memory formation during contextual fear conditioning. In the hippocampus 1 h after contextual fear conditioning, there were 675 demethylated genes and 613 hypermethylated genes. The consolidation of memory at 1 h after contextual fear conditioning was accompanied by the differential methylation of genes coding for ion channels, transcription factors, and constituents of the CREB and PKA signaling cascades, all of which have been shown to contribute to the early phases of learning and memory processes. These changes were transient in the hippocampal neurons, and almost none were present after 4 weeks. This also implies that the hypermethylated genes at 1 h then underwent active demethylation during the 4 weeks after contextual fear conditioning. Halder et al. [58], in addition, examined the anterior cingulate cortex, a brain region important for associative memory acquisition and maintenance of long-term memory. In the anterior cingulate cortex, at 1 h after contextual fear conditioning, there were 6250 differentially methylated genes, including 2423 demethylated genes. At 4 weeks after training 1223 differentially methylated genes persisted, including 118 demethylated genes. In addition, at 4 weeks after training they found 1700 differentially expressed genes in the anterior cingulate cortex. Their findings suggest that long-term memory (4 weeks) is associated with differential methylation of DNA and altered expression of more than a thousand genes in mouse neurons.

In 2017, Duke et al. [59], working with rats, studied neuron genomes in the hippocampus after contextual fear conditioning. At 24 h after contextual fear conditioning there were 2097 differentially methylated genes, with about 40% being demethylated. There were also 564 genes with upregulated expression and 1048 genes with downregulated expression. Hypermethylated regions overlapping differentially expressed genes were associated with decreased gene expression, consistent with the concept that cytosine methylation is often a mechanism for suppressing transcription. At 24 h after training, 9.2% of the genes in the rat genome of hippocampus neurons were differentially methylated. Gene Ontology term analysis was performed, and differentially expressed gene enrichment analysis revealed that many of the genes involved in synaptic functions were up-regulated 24 h after contextual fear conditioning in rats.

6.1 TET in learning and memory

In 2011, Guo et al. [26] were the first to show that TET1 is involved in neuronal activity-induced DNA demethylation and increased expression of memory-related genes in the mouse hippocampal dentate neurons. Demethylation of neuronal genes by TET1 appears to depend on TET1 being recruited to relevant genes. One mechanism of recruitment of TET appears to be by complexing with a specific "immediate early gene." The immediate early genes (IEGs) are a class of genes that are rapidly and transiently activated by a variety of signaling cascades and phosphorylation events, usually in a protein synthesis-independent manner, in response to neuronal activation [60]. ERG1 (Krox-24, Zif268) is an IEG product and is a neuronal activity-induced transcription factor. ERG1 appears to play an important role in learning and memory [60]. ERG1 is required specifically for the consolidation of long-term memory (while the related transcription factor ERG3 is primarily essential for short-term memory). As reviewed by Sun et al. [61], the short form of TET1, TET1s, is present in the brain. Sun et al. [61] experimentally showed that EGR1 and TET1s form a complex, independently of attachment to DNA. ERG1 undergoes rapid induction and appears to attach to binding sites at many genes upon neuronal activation. When ERG1 binds to a site, it is able to recruit a TET1s enzyme to that site. This allows TET1s to cause demethylation of a gene downstream of the binding site of EGR1, with upregulation of that gene's expression.

TET1 knockout mice [62] and *ERG1* knockout mice [63] are viable. Both have some developmental deficiencies [62, 63], and *TET1* knockouts [41, 43] and *ERG1* knockouts [64] each have some learning and memory deficiencies. Sun et al. [61] examined where differentially methylated regions occurred in the two types of knockout mice. Compared to wild-type mice, 322 and 2373 differentially methylated regions were identified in the brain frontal cortices (**Figure 6**) of *EGR1* knockout and *TET1* knockout mice respectively. There were 184 of these differentially methylated regions overlapping in the two types of knockout mice. This indicated that while ERG1 can bring TET1 to a DNA site to promote demethylation, TET1 is also brought too many other sites as well, presumably by other factors.

7. Conclusions

In evolutionary biology, the term exaptation refers to an evolutionary shift in the function of a trait over the course of natural selection [65]. For instance, a trait may evolve initially because it serves a particular function, but during the course of further evolution it may come to serve another function or an additional function. Such shifts in function are thought to be common in evolutionary history. As one example, bird feathers likely evolved initially for temperature regulation, and were later adapted for flight [65].

The idea that the function of a trait may shift during evolution was for many decades referred to as "preadaptation". However, this term suggests teleology in biology in conflict with natural selection and thus the term "preadaptation" has been replaced in the literature by "exaptation." This concept has recently been applied to the cognitive neurosciences [66]. It was proposed that substantial changes in function such as development of contemporary complex cognition including grammatical language, reading, writing and calculation abilities have occurred without evident changes in brain morphology over the past 150,000 years.

The evolutionary emergence of embryonic development also appears to have depended on an early exaptation. Enzymatic pathways that repair damage to the DNA genome likely existed very early in the history of life [67]. Processes that repair DNA, such as base excision repair, can also facilitate epigenetic modifications, particularly demethylation reactions, that alter gene expression and hence the function of cell lineages. Such epigenetic modifications play a central role in embryonic development including neurogenesis. Epigenetic alterations such as 5-methylcytosine are structurally similar to unwanted damages that are the primary target of DNA repair processes. Thus acquiring the new function of recognizing epigenetically methylated bases may have been enabled by this similarity. However, in the case of epigenetic demethylations, the effect of removing methyl groups and restoring the genome is to allow expression of genes that had been previously epigenetically silenced by methylation. Methylation and demethylation are reciprocal processes that appear to act coordinately to direct gene expression during embryonic development. DNA methylation reactions often cause silencing of gene expression, while demethylation reactions can reverse this process to allow expression. These mechanisms for controlling gene expression and the consequent facilitation of cell differentiation leading to embryonic development may have emerged in evolution as early as the origin of multicellular organisms more than 1 billion years ago [68].

Just as the evolutionary shift in the function of DNA repair appears to be central to the emergence of embryonic development and neurogenesis, this derived capability likely also gave rise to memory and learning. The molecular processes of epigenetic methylation and demethylation that underlie embryonic development also appear to underlie memory and learning. Thus the capacity for memory and learning may have evolved from a set of earlier epigenetic capabilities whose function was to promote embryonic reprogramming and neurogenesis.

In several neurodegenerative diseases epigenetic alterations appear to underlie characteristic features of the disease phenotype [69]. Proper functioning of the nervous system likely depends on DNA repair processes that not only restore DNA sequence information, but also facilitate normal gene expression by maintaining an appropriate set of epigenetic markers, particularly DNA methylation patterns. Understanding changes in DNA methylation patterns during early development and neurogenesis may contribute to the prevention or treatment of particular neurodegenerative diseases.

Parkinson disease patients treated with levodopa are subject to dyskinesia, a persistent behavioral sensitization that develops after levodopa exposure. Reorganization of DNA methylation patterns in the genome due to aberrant expression of DNA demethylation enzymes appears to have a pivotal role in the development of levodopa-induced dyskinesia [70]. Modification of DNA methylation is considered to be a promising novel therapeutic target for use in preventing or reversing dyskinetic behaviors [70]. Huntington's disease is a neurodegenerative disease that typically becomes apparent in midlife. This disease is associated with substantial changes in brain DNA methylation levels [71]. Aicardi-Goutieres syndrome (AGS) is a neurodegenerative condition characterized by early onset, often in infancy. Cells deficient in AGS proteins display a substantial 5–20% reduction in genomic methylation levels overall, and this reduction is distributed widely in the genome [72]. The fragile X syndrome is a prevalent form of mental retardation. This condition is caused by loss of expression of the *FMR1* gene, usually due to expansion of a CGG repeat sequence (>200 repeats) in the first exon of *FMR1*. This sequence expansion leads to abnormal methylation of the promoter region that then causes transcriptional silencing of the *FMR1* gene and an absence of the fragile X mental retardation protein [73]. Several studies have described methylation alterations in various regions of the brain in Alzheimer's disease, as reviewed by Yokoyama et al. [74]. The results of these studies, so far, appear to be somewhat contradictory and additional studies will be needed to provide clear conclusions. These various studies of DNA methylation alterations are still at an early stage, but nevertheless suggest that as our basic understanding of how epigenetic DNA methylation patterns influence neurodegenerative disease advances, this understanding will contribute to disease prevention and treatment.

Author details

Carol Bernstein* and Harris Bernstein
University of Arizona, Tucson, Arizona USA

*Address all correspondence to: bernstein324@yahoo.com

References

[1] Schopf JW. Fossil evidence of Archaean life. Philosophical Transactions of the Royal Society of London. Series B, Biological Sciences. 2006;**361**(1470):869-885. DOI: 10.1098/rstb.2006.1834

[2] De Bont R, van Larebeke N. Endogenous DNA damage in humans: A review of quantitative data. Mutagenesis. 2004;**19**(3):169-185. DOI: 10.1093/mutage/geh025

[3] https://www.mdanderson.org/documents/Labs/Wood-Laboratory/human-dna-repair-genes.html

[4] Bernstein C, Bernstein H, Payne CM, Garewal H. DNA repair/pro-apoptotic dual-role proteins in five major DNA repair pathways: Fail-safe protection against carcinogenesis. Mutation Research. 2002;**511**(2):145-178. DOI: 10.1016/s1383-5742(02)00009-1

[5] Valavanidis A, Vlachogianni T, Fiotakis K, Loridas S. Pulmonary oxidative stress, inflammation and cancer: Respirable particulate matter, fibrous dusts and ozone as major causes of lung carcinogenesis through reactive oxygen species mechanisms. International Journal of Environmental Research and Public Health. 2013;**10**(9):3886-3907. DOI: 10.3390/ijerph10093886

[6] Covarrubias L, Hernández-García D, Schnabel D, Salas-Vidal E, Castro-Obregón S. Function of reactive oxygen species during animal development: Passive or active? Developmental Biology. 2008;**320**(1):1-11. DOI: 10.1016/j.ydbio.2008.04.041

[7] Ufer C, Wang CC, Borchert A, Heydeck D, Kuhn H. Redox control in mammalian embryo development. Antioxidants & Redox Signaling. 2010;**13**(6):833-875. DOI: 10.1089/ars.2009.3044

[8] Massaad CA, Klann E. Reactive oxygen species in the regulation of synaptic plasticity and memory. Antioxidants & Redox Signaling. 2011;**14**(10):2013-2054. DOI: 10.1089/ars.2010.3208

[9] Beckhauser TF, Francis-Oliveira J, De Pasquale R. Reactive oxygen species: Physiological and physiopathological effects on synaptic plasticity. Journal of Experimental Neuroscience. 2016;**10**(Suppl 1):23-48. DOI: 10.4137/JEN.S39887

[10] Wu X, Li G, Xie R. Decoding the role of TET family dioxygenases in lineage specification. Epigenetics & Chromatin. 2018;**11**(1):58. DOI: 10.1186/s13072-018-0228-7

[11] Alaghband Y, Bredy TW, Wood MA. The role of active DNA demethylation and Tet enzyme function in memory formation and cocaine action. Neuroscience Letters. 2016;**625**:40-46. DOI: 10.1016/j.neulet.2016.01.023

[12] Messerschmidt DM, Knowles BB, Solter D. DNA methylation dynamics during epigenetic reprogramming in the germline and preimplantation embryos. Genes & Development. 2014;**28**(8):812-828. DOI: 10.1101/gad.234294.113

[13] Santos F, Hendrich B, Reik W, Dean W. Dynamic reprogramming of DNA methylation in the early mouse embryo. Developmental Biology. 2002;**241**(1):172-182. DOI: 10.1006/dbio.2001.0501

[14] Zeng Y, Chen T. DNA methylation reprogramming during mammalian development. Genes (Basel). 2019;**10**(4):pii: E257. DOI: 10.3390/genes10040257

[15] S Sridhar17. Own Work. CC BY-SA 4.0. Available from: https://commons.wikimedia.org/wiki/File:CytosineMethylation.png

[16] Jang HS, Shin WJ, Lee JE, Do JT. CpG and non-CpG methylation in epigenetic gene regulation and brain function. Genes (Basel). 2017;**8**(6):pii: E148. DOI: 10.3390/genes8060148

[17] Luo Y, Lu X, Xie H. Dynamic Alu methylation during normal development, aging, and tumorigenesis. BioMed Research International. 2014;**2014**:784706. DOI: 10.1155/2014/784706

[18] Edwards JR, O'Donnell AH, Rollins RA, Peckham HE, Lee C, Milekic MH, et al. Chromatin and sequence features that define the fine and gross structure of genomic methylation patterns. Genome Research. 2010;**20**(7):972-980. DOI: 10.1101/gr.101535.109

[19] Ming X, Matter B, Song M, Veliath E, Shanley R, Jones R, et al. Mapping structurally defined guanine oxidation products along DNA duplexes: Influence of local sequence context and endogenous cytosine methylation. Journal of the American Chemical Society. 2014;**136**(11):4223-4235. DOI: 10.1021/ja411636j

[20] Ziller MJ, Müller F, Liao J, Zhang Y, Gu H, Bock C, et al. Genomic distribution and inter-sample variation of non-CpG methylation across human cell types. PLoS Genetics. 2011;**7**(12):e1002389. DOI: 10.1371/journal.pgen.1002389

[21] Guénet JL. The mouse genome. Genome Research. 2005;**15**(12):1729-1740. DOI: 10.1101/gr.3728305

[22] Haaf T. The battle of the sexes after fertilization: Behaviour of paternal and maternal chromosomes in the early mammalian embryo. Chromosome Research. 2001;**9**(4):263-271. PMID: 11419791

[23] Bayraktar G, Kreutz MR. The role of activity-dependent DNA demethylation in the adult brain and in neurological disorders. Frontiers in Molecular Neuroscience. 2018;**11**:169. DOI: 10.3389/fnmol.2018.00169

[24] Bernstein0275. Own Work. CC BY-SA 4.0. Available from: https://commons.wikimedia.org/wiki/File:Demethylation_of_5-methylcytosine.svg

[25] Zou J, Wang C, Ma X, Wang E, Peng G. APOBEC3B, a molecular driver of mutagenesis in human cancers. Cell & Bioscience. 2017;**7**:29. DOI: 10.1186/s13578-017-0156-4

[26] Guo JU, Su Y, Zhong C, Ming GL, Song H. Hydroxylation of 5-methylcytosine by TET1 promotes active DNA demethylation in the adult brain. Cell. 2011;**145**(3):423-434. DOI: 10.1016/j.cell.2011.03.022

[27] Jin SG, Zhang ZM, Dunwell TL, Harter MR, Wu X, Johnson J, et al. Tet3 reads 5-carboxylcytosine through its CXXC domain and is a potential guardian against neurodegeneration. Cell Reports. 2016;**14**(3):493-505. DOI: 10.1016/j.celrep.2015.12.044

[28] Melamed P, Yosefzon Y, David C, Tsukerman A, Pnueli L. Tet enzymes, variants, and differential effects on function. Frontiers in Cell and Development Biology. 2018;**6**:22. DOI: 10.3389/fcell.2018.00022

[29] Wossidlo M, Nakamura T, Lepikhov K, Marques CJ, Zakhartchenko V, Boiani M, et al. 5-Hydroxymethylcytosine in the mammalian zygote is linked with epigenetic reprogramming. Nature Communications. 2011;**2**:241. DOI: 10.1038/ncomms1240

[30] Zhang W, Xia W, Wang Q, Towers AJ, Chen J, Gao R, et al. Isoform switch of TET1 regulates DNA demethylation and mouse development. Molecular Cell. 2016;**64**(6):1062-1073. DOI: 10.1016/j.molcel.2016.10.030

[31] Deplus R, Delatte B, Schwinn MK, Defrance M, Méndez J, Murphy N, et al. TET2 and TET3 regulate GlcNAcylation and H3K4 methylation through OGT and SET1/COMPASS. The EMBO Journal. 2013;**32**(5):645-655. DOI: 10.1038/emboj.2012.357

[32] Swenberg JA, Lu K, Moeller BC, Gao L, Upton PB, Nakamura J, et al. Endogenous versus exogenous DNA adducts: Their role in carcinogenesis, epidemiology, and risk assessment. Toxicological Sciences. 2011;**120**(Suppl 1):S130-S145. DOI: 10.1093/toxsci/kfq371

[33] Abdou I, Poirier GG, Hendzel MJ, Weinfeld M. DNA ligase III acts as a DNA strand break sensor in the cellular orchestration of DNA strand break repair. Nucleic Acids Research. 2015;**43**(2):875-892. DOI: 10.1093/nar/gku1307

[34] Sakumi K, Tominaga Y, Furuichi M, Xu P, Tsuzuki T, Sekiguchi M, et al. Ogg1 knockout-associated lung tumorigenesis and its suppression by Mth1 gene disruption. Cancer Research. 2003;**63**(5):902-905. PMID: 12615700

[35] Hamilton ML, Guo Z, Fuller CD, Van Remmen H, Ward WF, Austad SN, et al. A reliable assessment of 8-oxo-2-deoxyguanosine levels in nuclear and mitochondrial DNA using the sodium iodide method to isolate DNA. Nucleic Acids Research. 2001;**29**(10):2117-2126. DOI: 10.1093/nar/29.10.2117

[36] Zhou X, Zhuang Z, Wang W, He L, Wu H, Cao Y, et al. OGG1 is essential in oxidative stress induced DNA demethylation. Cellular Signalling. 2016;**28**(9):1163-1171. DOI: 10.1016/j.cellsig.2016.05.021

[37] Bernstein0275. Own Work. CC BY-SA 4.0. Available from: https://en.wikipedia.org/wiki/Epigenetics_in_learning_and_memory#/media/File:Initiation_of_DNA_demethylation_at_a_CpG_site.svg

[38] Klungland A, Rosewell I, Hollenbach S, Larsen E, Daly G, Epe B, et al. Accumulation of premutagenic DNA lesions in mice defective in removal of oxidative base damage. Proceedings of the National Academy of Sciences of the United States of America. 1999;**96**(23):13300-13305. DOI: 10.1073/pnas.96.23.13300

[39] Miller-Pinsler L, Pinto DJ, Wells PG. Oxidative DNA damage in the in utero initiation of postnatal neurodevelopmental deficits by normal fetal and ethanol-enhanced oxidative stress in oxoguanine glycosylase 1 knockout mice. Free Radical Biology & Medicine. 2015;**78**:23-29. DOI: 10.1016/j.freeradbiomed.2014.09.026

[40] Seifermann M, Epe B. Oxidatively generated base modifications in DNA: Not only carcinogenic risk factor but also regulatory mark? Free Radical Biology & Medicine. 2017;**107**:258-265. DOI: 10.1016/j.freeradbiomed.2016.11.018

[41] Zhang RR, Cui QY, Murai K, Lim YC, Smith ZD, Jin S, et al. Tet1 regulates adult hippocampal neurogenesis and cognition. Cell Stem Cell. 2013;**13**(2):237-245. DOI: 10.1016/j.stem.2013.05.006

[42] Rudenko A, Dawlaty MM, Seo J, Cheng AW, Meng J, Le T, et al. Tet1 is critical for neuronal activity-regulated gene expression and memory extinction. Neuron. 2013;**79**(6):1109-1122. DOI: 10.1016/j.neuron.2013.08.003

[43] Kaas GA, Zhong C, Eason DE, Ross DL, Vachhani RV, Ming GL, et al. TET1 controls CNS 5-methylcytosine hydroxylation, active DNA demethylation, gene transcription, and memory formation. Neuron. 2013;**79**(6):1086-1093. DOI: 10.1016/j.neuron.2013.08.032

[44] Gu TP, Guo F, Yang H, Wu HP, Xu GF, Liu W, et al. The role of Tet3 DNA dioxygenase in epigenetic reprogramming by oocytes. Nature. 2011;**477**(7366):606-610. DOI: 10.1038/nature10443

[45] Zhao H, Chen T. Tet family of 5-methylcytosine dioxygenases in mammalian development. Journal of Human Genetics. 2013;**58**(7):421-427. DOI: 10.1038/jhg.2013.63

[46] Wang Z, Tang B, He Y, Jin P. DNA methylation dynamics in neurogenesis. Epigenomics. 2016;**8**(3):401-414. DOI: 10.2217/epi.15.119

[47] Pilz GA, Bottes S, Betizeau M, Jörg DJ, Carta S, Simons BD, et al. Live imaging of neurogenesis in the adult mouse hippocampus. Science. 2018;**359**(6376):658-662. DOI: 10.1126/science.aao5056

[48] Alberini CM. Long-term memories: The good, the bad, and the ugly. Cerebrum. 2010;**2010**:21. PMID: 23447766

[49] Griffith JS, Mahler HR. DNA ticketing theory of memory. Nature. 1969;**223**(5206):580-582. DOI: 10.1038/223580a0

[50] Holliday R. Is there an epigenetic component in long-term memory? Journal of Theoretical Biology. 1999;**200**(3):339-341. DOI: 10.1006/jtbi.1999.0995

[51] Kim JJ, Jung MW. Neural circuits and mechanisms involved in Pavlovian fear conditioning: A critical review. Neuroscience and Biobehavioral Reviews. 2006;**30**(2):188-202. PMID: 16120461

[52] https://commons.wikimedia.org/wiki/File:Brain_regions_involved_in_memory_formation.jpg

[53] Frankland PW, Bontempi B, Talton LE, Kaczmarek L, Silva AJ. The involvement of the anterior cingulate cortex in remote contextual fear memory. Science. 2004;**304**(5672):881-883. DOI: 10.1126/science.1094804

[54] Baumgärtel K, Mansuy IM. Neural functions of calcineurin in synaptic plasticity and memory. Learning & Memory. 2012;**19**(9):375-384. DOI: 10.1101/lm.027201.112

[55] Miller CA, Gavin CF, White JA, Parrish RR, Honasoge A, Yancey CR, et al. Cortical DNA methylation maintains remote memory. Nature Neuroscience. 2010;**13**(6):664-666. DOI: 10.1038/nn.2560

[56] Cunha C, Brambilla R, Thomas KL. A simple role for BDNF in learning and memory? Frontiers in Molecular Neuroscience. 2010;**3**:1. DOI: 10.3389/neuro.02.001.2010

[57] Lubin FD, Roth TL, Sweatt JD. Epigenetic regulation of BDNF gene transcription in the consolidation of fear memory. The Journal of Neuroscience. 2008;**28**(42):10576-10586. DOI: 10.1523/JNEUROSCI.1786-08.2008

[58] Halder R, Hennion M, Vidal RO, Shomroni O, Rahman RU, Rajput A, et al. DNA methylation changes in plasticity genes accompany the formation and maintenance of memory. Nature Neuroscience. 2016;**19**(1):102-110. DOI: 10.1038/nn.4194

[59] Duke CG, Kennedy AJ, Gavin CF, Day JJ, Sweatt JD. Experience-dependent epigenomic reorganization in the hippocampus. Learning & Memory.

2017;**24**(7):278-288. DOI: 10.1101/lm.045112.117

[60] Poirier R, Cheval H, Mailhes C, Garel S, Charnay P, Davis S, et al. Distinct functions of egr gene family members in cognitive processes. Frontiers in Neuroscience. 2008;**2**(1):47-55. DOI: 10.3389/neuro.01.002.2008

[61] Sun Z, Xu X, He J, Murray A, Sun MA, Wei X, et al. EGR1 recruits TET1 to shape the brain methylome during development and upon neuronal activity. Nature Communications. 2019;**10**(1):3892. DOI: 10.1038/s41467-019-11905-3

[62] Dawlaty MM, Ganz K, Powell BE, Hu YC, Markoulaki S, Cheng AW, et al. Tet1 is dispensable for maintaining pluripotency and its loss is compatible with embryonic and postnatal development. Cell Stem Cell. 2011;**9**(2):166-175. DOI: 10.1016/j.stem.2011.07.010

[63] Topilko P, Schneider-Maunoury S, Levi G, Trembleau A, Gourdji D, Driancourt MA, et al. Multiple pituitary and ovarian defects in Krox-24 (NGFI-A, Egr-1)-targeted mice. Molecular Endocrinology. 1998;**12**(1):107-122. DOI: 10.1210/mend.12.1.0049

[64] Jones MW, Errington ML, French PJ, Fine A, Bliss TV, Garel S, et al. A requirement for the immediate early gene Zif268 in the expression of late LTP and long-term memories. Nature Neuroscience. 2001;**4**(3):289-296. DOI: 10.1038/85138

[65] Gould SJ, Vrba ES. Exaptation—A missing term in the science of form. Paleobiology. 1982;**8**(1):4-15. DOI: 10.1017/S0094837300004310

[66] Ardila A. The evolutionary concept of "preadaptation" applied to cognitive neurosciences. Frontiers in Neuroscience. 2016;**10**:103. DOI: 10.3389/fnins.2016.00103

[67] DiRuggiero J, Brown JR, Bogert AP, Robb FT. DNA repair systems in archaea: Mementos from the last universal common ancestor? Journal of Molecular Evolution. 1999;**49**(4):474-484. DOI: 10.1007/pl00006570

[68] Strother PK, Battison L, Brasier MD, Wellman CH. Earth's earliest non-marine eukaryotes. Nature. 2011;**473**(7348):505-509. DOI: 10.1038/nature09943

[69] Bernstein C, Bernstein H. The role of DNA repair and the epigenetic markers left after repair in neurologic functions, including memory and learning. In: Mognato M, editor. DNA Repair—An Update. Rijeka, Croatia: IntechOpen; 2018. pp. 105-126. DOI: 10.5772/intechopen.83477

[70] Figge DA, Eskow Jaunarajs KL, Standaert DG. Dynamic DNA methylation regulates levodopa-induced dyskinesia. The Journal of Neuroscience. 2016;**36**:6514-6524. DOI: 10.1523/JNEUROSCI.0683-16.2016

[71] Horvath S, Langfelder P, Kwak S, Aaronson J, Rosinski J, Vogt TF, et al. Huntington's disease accelerates epigenetic aging of human brain and disrupts DNA methylation levels. Aging (Albany NY). 2016;**8**:1485-1512. DOI: 10.18632/aging.101005

[72] Lim YW, Sanz LA, Xu X, Hartone SR, Chetin F. Genome-wide DNA hypomethylation and RNA:DNA hybrid accumulation in Aicardi-Goutieres syndrome. eLife. 2015;**4**:08007. DOI: 10.7554/eLife.08007

[73] Serrano M. Epigenetic cerebellar diseases. Handbook of Clinical Neurology. 2018;**155**:227-244. DOI: 10.1016/B978-0-444-64189-2.00015-9

[74] Yokoyama AS, Rutledge JC, Medici V. DNA methylation alterations in Alzheimer's disease. Environmental Epigenetics. 2017;**3**:dvx008. DOI: 10.1093/eep/dvx008

Chapter 3

Global DNA Methylation as a Potential Underlying Mechanism of Congenital Disease Development

Aleksandra Stanković

Abstract

During the last decade, quantitative measurement of the methylation status in white blood cells (WBCs) has been used as a potential biomarker in a variety of diseases. Long interspersed nucleotide element-1 (LINE-1) has been used widely as a surrogate marker of global DNA methylation. Altered maternal DNA methylation is suggested to be an underlying mechanism in the trisomy 21 and the development of birth defects, including congenital heart defects (CHDs). The molecular mechanisms that underlie the epigenetic regulation of gene transcription are independent of DNA sequence, but they do depend on environmental stimuli, which are especially important in fetal development in utero environment. Folic acid deficiency and genetic variations of folate pathway genes, such as the methylenetetrahydrofolate reductase gene (*MTHFR*), are foremost among these maternal risk factors. Also, there are exogenous risk factors (cigarette smoking, alcohol intake, medication use, periconceptional maternal supplementation, body mass index, and dietary habits) with impact on maternal LINE-1 methylation. *MTHFR* C677T genotype/diet and other environmental factors as significant predictors of LINE-1 DNA methylation in regard to congenital diseases will be discussed in the chapter.

Keywords: DNA methylation, LINE-1, congenital anomaly, development, nutrition, folate intake, genotype

1. Introduction

According to WHO, congenital anomalies (CAs) are birth defects that can be defined as structural or functional malformations [1]. CAs occur during intrauterine life and can be identified prenatally or at birth or later in infancy. CAs are important causes of infant and childhood deaths and chronic illness/disability. Long-term disability may have significant impacts on patients, families, healthcare systems, and societies. Some CAs can be prevented by adequate intake of folic acid (FA) through fortification of staple foods or supplementation. Among severe congenital anomalies, the most common ones are congenital heart defects (CHDs), neural tube defects (NTD), and Down syndrome (DS). For renal dysplasia an increasing trend was observed recently in Europe [2]. Several factors have been proposed to have a significant role in the development of CAs: one or more genes;

infectious, maternal diabetes or obesity; and nutritional and environmental factors [2]. Identification of the exact cause/causes recently became even more complicated with addition of new factors. Epigenetic factors, as it is DNA methylation, have been shown to have an impact on the gene expression, through modulation by nutrition or environmental stimuli that occur during intrauterine development, but could even be a consequence of maternal or paternal lifestyle factors. Altered DNA methylation was suggested to be an underlying mechanism in the development of CAs, CHDs, NTD, congenital anomaly of the kidney and urinary tract (CAKUT), and autism spectrum disorders (ASD) and in imprinting genetic disorders [3–12]. Congenital heart defects (CHDs) are the most common birth defects in humans with a prevalence of 0.8% [13, 14]. Only about 15–20% of CHDs can be attributed to known causes, whereas chromosomal abnormalities occur in 5–10% of cases [14]. The highest association with major heart abnormalities is observed in DS [15]. CHDs are reported in approximately 40% of DS cases, typically involving septal defects such as atrial septal defect, ventricular septal defect, and complete atrio-ventricular canal [16]. The etiology of most CHDs remains largely unknown, but it is considered to involve multiple genetic, epigenetic, environmental, and lifestyle factors [13, 14, 17]. Risk factors, including aging, body mass index (BMI), cigarette smoking, alcohol intake, folate deficiency, MTHFR polymorphisms, and hyper-homocysteinemia, have been proposed to be the modulators of DNA methylation patterns [3–6, 18–20]. Maternal intrauterine milieu, such as maternal environment during pregnancy (hypoxia, stress, obesity, diabetes, toxins, altered nutrition, inflammation, and reduced utero-placental blood flow) could affect fetal methylation programming, thereby affecting fetal growth and the lifelong health of the fetus [21, 22]. It was reported that the maternal LINE-1 hypomethylation is linked with the increased risk for non-syndromic CHD, particularly septal defects [4, 5].

2. DNA gene-specific methylation and global DNA methylation

DNA methylation is a key factor of the epigenetic machinery that is responsible for regulating gene expression and, therefore, cell function. This component is one of the most important in mammalian embryonic development, differentiation, and many of congenital and complex diseases [3–6, 23–25]. The DNA methylation has nonrandom, well-regulated, and tissue-specific patterns [26]. Abnormal gene-specific demethylation and global hypomethylation (involving repeat sequences throughout the genome) can potentially lead to overexpression of genes and activation of transposable elements contributing to disease. Regulation of gene expression through methylated or unmethylated human genome can exist at approximately 3×10^7 CpG short sequences of 5–10 CpG dinucleotides [27, 28].

DNA methylation is required in many processes such as X chromosome inactivation, imprinting, embryogenesis, gametogenesis, and silencing of repetitive DNA elements [29]. It refers to the covalent addition of a methyl group to the cytosine located at the 5′-position to guanosine in a CpG dinucleotide, catalyzed by the activity of three DNA methyltransferases (DNMTs) [30]. Recent findings of tissue-specific expression of ten-eleven translocation (TET) proteins revealed that this epigenetic event is not irreversible and, even more, TET was shown to be able to modify methylcytosine and potentially erase DNA methylation [31].

Each of the three DNMT genes was found to be mutated in specific and diverse human syndromes [32]. DNA methylation is required to protect chromosomal integrity, by preventing reactivation of endoparasitic sequences that cause chromosomal instability, translocations, increased mutation events, loss of imprinting, and gene disruption [29]. Genome-wide methylation profiling has recently become possible and revealed genes of interest that were enriched in multiple biological processes involved in fetal development [3], and specific hypermethylation was linked to gene silencing in some pediatric disorders [33, 34]. Moreover, epigenetic mechanisms including parent of origin-specific DNA methylation include genomic imprinting as restriction of gene expression [35]. Moreover, imprinting in embryos was found to be parentally sex-specific, and this effect could be more complex than previously suggested [36]. Hypomethylation of imprinted loci (HIL) throughout the genome was observed in patients with imprinted disorders. Among approximately 70 known imprinted genes, there are some that are causing disorders affecting growth, including one in the DS critical region [35]. Aberrant methylation in four maternally methylated regions was observed at whole genome methylation analysis. However, methylation of a CpG island does not necessarily lead to gene silencing. For example, the gene for telomerase has been shown to be activated by methylation [37]. Telomerases are crucial elements in maintaining cell life, could possibly reverse an aging mechanism, and rejuvenate cell viability. Enzyme telomerase modulates elongation of telomeres, by adding repeating DNA sequences to the ends of the chromosomes, and telomere serves as a bioprotective mechanism of chromosome attrition at each cell division [38]. Telomeres could become too short to allow replication or dysfunctional in some congenital disease which may lead to chromosome instability or cell death [39]. Besides DNA coding region, studies have shown that DNA methylation of noncoding DNA plays an important role in modulating structure and dynamics of chromatin, as well as many other chromatin-dependent processes and their associated biological functions [27].

2.1 LINE-1 DNA methylation

Gene-specific DNA methylation analysis does not provide a global picture of DNA methylation changes within a genome. Global DNA hypomethylation occurs mainly at heavily methylated noncoding regions of DNA, particularly repeat sequences and transposable elements [40, 41].

In humans, nearly 80% CpG islands occur in transposon-derived sequences, throughout the genome, such as long interspersed nuclear elements (LINEs) and short interspersed nuclear elements (SINEs) [42]. LINE-1 is the largest member of the LINE family with more than 500,000 copies comprising approximately 17% of the genome [43]. CpG islands within LINE-1 sequences and their methylation levels correlate with the global genomic DNA methylation level [44, 45], so LINE-1 methylation has been widely used as a surrogate marker of global genomic DNA methylation [46], and methylation status of LINE-1 in white blood cells (WBC) is a potential biomarker in a variety of diseases [4, 45–48] in research on cancer, cardiovascular, neurodegenerative, and CAs [3–6, 48–51]. Human genome has on average 80–100 active LINE-1, and it has been estimated that new LINE-1 insertion in genome occurs in at least 1 in every 50 humans within a parental germ cell or during early fetal development [40]. Thus, LINE-1 hypomethylation in the parental germline, along with altered miRNA expression, might also significantly affect genome stability during the fetal development [52, 53].

3. DNA methylation during gametogenesis and embryogenesis

DNA methylation changes are particularly dynamic in gametogenesis and early embryogenesis. During the course of mammalian differentiation and development, DNA methylation undergoes remodeling to eventually generate the cell type-specific methylation patterns, found in somatic cells of adults. During the gametogenesis, DNA is demethylated within each developing germ cell and then remethylated/reset to the methylation patterns specific to gametogenesis. The differentially methylated regions (DMRs) are sperm and egg specific [54, 55]. This process establishing the specific methylation of imprinted loci before fertilization, as well as other non-imprinted loci, may also be subject to at least partial erasure of methylation during gametogenesis [56–58]. The zygotic DNA demethylation after fertilization in mouse embryogenesis affects parental genome on a genome-wide level including single gene loci and repetitive elements. The maternal genome-wide methylation is unaffected [59]. This process changes the methylation patterns of the gametes and establishes the DNA methylation patterns found in somatic differentiated cells in adults through induced expression of DNMT and de novo methylation of genome in post-implantation mouse embryos [60–62]. It has been shown that in small studies of human embryos, there is a demethylation process at the 4-cell stage followed by remethylation at late morula [63]. Even more, expression patterns of DNMTs after cryopreservation of human embryos could be disturbed and could have long-term developmental consequences [64] that suggest the importance of DNA methylation program maintenance during development. Periods during gametogenesis and embryogenesis may also present windows of opportunity for environmental influences on DNA methylation pattern. The DMRs are established during gametogenesis at imprinted and non-imprinted loci and are susceptible to environmental factors [65, 66]. LINE-1 methylation in sperm could be a risk marker of infertility in man at nicotine/alcohol exposure [67]. It is also possible to alter DNA methylation levels and patterns within intact mammalian cells by treatment with various chemical inhibitors, DNA-demethylating drugs, which have recently been introduced as potential therapeutic agents for the treatment of human diseases, particularly myelodysplastic syndromes [68].

The dynamic reprogramming and other epigenetic patterns which could affect normal patterns of gene expression/genome stability during development could lead to an increased risk of CAs or complex diseases later in life [65–67].

4. LINE-1 DNA methylation and environmental influences (e.g., diet and nutrition)

Previous research was focused on the effect of specific foods on the DNA methylation process, but there is currently growing interest in determining how dietary patterns may affect global and local DNA methylation in humans. There are some studies that suggest that frequent use of vegetables and/or fruits decreased the risk of LINE-1 hypomethylation [69–71]. Biological explanation could be in beneficial modulation of pathways involved in epigenetic mechanisms by intake of high variety of nutritive and bioactive substances included in fruit- and vegetable-rich food. These components were polyphenols; flavonoids; carotenoids; folates; vitamins C, E, and A; minerals; and fibers [72, 73]. As it is known that many crucial cellular processes depend on folate, including DNA methylation [74], low folate intake in daily food could be supplemented by synthetic form as folic acid (FA) and through fortification programs [75]. Even more, harmful effect of particulate matter exposure on LINE-1 methylation level could be counteracted by healthy food consumption such as Mediterranean diet [76]. Also, fatty acids can modify DNA methylation in vitro, but limited information is available from human studies. Some studies observed that intake of vegetable oil/dietary fat seemed to be negatively correlated with LINE-1 methylation [69, 77]. Others show no changes in methylation profile after supplementation with grape seed flavanols [78]. The interindividual variation in blood cell DNA methylation in interventional studies, which are usually rather small, demands studies with larger sample size to avoid masking the possibly subtle changes in DNA methylation in response to dietary factors.

4.1 Methylenetetrahydrofolate reductase (MTHFR), folate metabolism, and its role in DNA methylation

Folate can be a limiting factor in many biological reactions. The methylene tetrahydrofolate reductase (MTHFR) is an enzyme important for the folate metabolism which is in the basis of the DNA, RNA, and protein methylation. Genomic DNA methylation directly correlates with folate status and inversely with plasma homocysteine (tHcy) levels [79–82]. The one-carbon pathway and thus DNA methylation function under tight regulatory controls. S-Adenosyl methionine (SAM) is the major regulator of folate-dependent Hcy remethylation because it is a potent inhibitor of MTHFR. When the SAM concentration is high, MTHFR is inhibited and hence remethylation of homocysteine. Conversely, if SAM concentrations are low, remethylation of homocysteine is favored. Hyperhomocysteinemia is an emerging risk factor for various cardiovascular diseases, and, with the increasing significance of this genetic variant in the view of morbidity and mortality impact on the patients, further prevention strategies and nutritional recommendations with the supplementation of folate would be necessary as part of future health education. Other essential nutrients that are naturally present in some foods or as dietary supplement, like vitamin B_6, B_{12}, B_2, and choline, are necessary in addition to folate to maintain DNA methylation [83]. It is also recognized that S-adenosylhomocysteine (SAH) functions as a potent product inhibitor of SAM-dependent methyltransferases [84]. For this reason, continual hydrolysis of SAH to homocysteine is important for DNA methylation [85]. Plasma homocysteine elevation has been associated with increased concentration of SAH, and increased SAH was in correlation with global DNA hypomethylation [86]. Methionine is the substrate for SAM, a cofactor and methyl group donor for numerous methylation reactions including the methylation of DNA, RNA, and histones [87]. A number of SAM-dependent reactions have regulatory roles by affecting both, genome stability and gene transcription [88].

4.2 Epigenetic, genetic, and nutrigenomic risk factors for congenital diseases: DNA methylation, global DNA methylation, miRNA, MTHFR polymorphism, and low folate status

Low folate status (as defined by various measures including blood folate concentrations, folate intake, and/or FA intake) has been associated with an increased risk of cardiovascular disease, cancers, CAs, CHD, and NTD [5, 6, 89–94]. Also, this deficiency is clearly detrimental to the embryo and shows possible longer-term risks of diabetes or other health outcomes and health problems associated with child mortality and morbidity [95]. Periconceptional supplementation of FA also reduces the risk of congenital heart diseases (previous ref) and preterm birth and low birth weight [96, 97]. The prevalence of neural tube defects (NTDs) has been significantly lowered in more than 70 countries worldwide by applying fortification with FA, but in all European governments there is still an issue with FA fortification of centrally processed and widely eaten foods in prevention of unwanted birth outcomes [98]. The mechanisms by which low folate status contributes to these disorders have not been understood completely but, to a certain extent, could be explained by different molecular pathways. Folate depletion could be a destabilizing factor during DNA replication. If inadequate folate availability is present during cell division, the production of thymidine could be compromised and may be substituted in the DNA sequence by uracil. This mutagenic event may trigger the defect in an effort to repair DNA and increase the frequency of chromosomal breaks [90]. Low FA in tissue culture has been shown to result in the formation of micronuclei (chromosome breakage) and that the presence of MTHFR C677T polymorphism (TT genotype) increases the micronuclei formation, under the low folate conditions [99]. This MTHFR polymorphism was associated with various diseases, and allele frequencies vary depending on ethnicity (reviewed in [100]). This gene is mapped on chromosome 1 (1p36.6), and the genetic variant assigned as C677T (rs1801133) is located in exon 4 in this gene. This polymorphism results in the conversion at codon 222, valine to alanine. Carriers of the T allele have lower enzyme activity [101]. The MTHFR 677TT homozygous subjects have higher homocysteine levels than the normal, non-mutated controls. To date, most studies have shown that the *MTHFR* C677T genotype is related to biomarkers, such as serum folate, tHcy concentration, and folate intake. Elevated blood tHcy is a well-recognized and modifiable risk factor for cerebral and cardiovascular disease [101, 102]. Reduction of the enzyme activity leads to elevated Hcy concentrations [103]. The TT genotype has been associated with elevated tHcy levels in populations with low folate intake [104]. Previous tHcy-lowering trials have not considered whether and to what extent these factors could modify the efficacy of folic acid (FA) treatment. In some countries with folate fortification like America, Australia, and New Zealand, the effect of TT genotype is not so obvious like in Asia region where folate intake is low [94]. In those who are homozygous for the mutation (TT genotype), enzyme function is only 30% of normal, and data provide evidence that nutrition can counteract genetic susceptibility. Recently, large, randomized trial in a population without mandatory FA fortification demonstrated that the adverse effect of the TT genotype on tHcy levels can be ameliorated by raising serum folate levels above the threshold (15 ng/mg or 34 nmol/L) via FA treatment and it provides new evidence to support a personalized FA treatment [94]. The gene-nutrient interaction between MTHFR C677T variant and folate status was also observed on the risk of anencephaly. Mothers with 677TT genotype with serum folate levels in the upper tercile (>14.1 ng/ml) had a 95% lower risk to have a child with anencephaly than mothers with serum folate levels in the first and second terciles [92]. Results about DS and MTHFRC677T polymorphism as a risk factor of its occurrence are still conflicting. The recent meta-analysis suggested that MTHFR 677T is a major risk factor for DS birth [105], while previous smaller studies did not recognize such risk [106, 107]. Studies performed analyzing peripheral lymphocytes of women with DS offspring revealed several markers of global genome instability, including an increased frequency of micronuclei, shorter telomeres, and impaired DNA methylation at MTHFR promoter [108, 109]. Hypermethylation of MTHFR promoter may lead to CHD in DS subjects [109]. Functional inactivation of MTHFR gene expression could be a mechanism of impaired folate metabolism, which is known to play a role in chromosomal breakage, abnormal chromosomal segregation, and genomic instability and therefore a developmental defect in the CHD in DS. Another suggested mechanism is lower LINE-1 methylation, the surrogate marker for global methylation levels, in young mothers of DS compared to controls, suggesting the possibility of impaired DNA methylation causing maternally derived trisomy 21 [6]. Also, there is evidence from intervention studies of the effects of dietary factors, where FA was the most common intervention agent (33%). Meta-regression analysis showed that the dose of supplementary FA was the only identified factor ($p < 0.001$) showing a positive relationship with DNA methylation patterns in humans [93]. MTHFR genotype-dependent association between lower global DNA methylation and lower plasma folate concentration was detected in observational studies in healthy subjects [81, 82, 110]. Global DNA methylation at maternal front ($p = 0.04$) and hypomethylation of MTHFR gene at fetal front ($p = 0.001$) might be a characteristic of preeclampsia [111]. The combination of MTHFR C677T genotype and diet significantly influenced global DNA methylation in mothers with DS children. The lowest values of global DNA methylation were observed in mothers with MTHFR 677 CT+TT genotype and low dietary folate [6]. Even more, recently the association between maternal LINE-1 methylation and the occurrence of CHD in children with DS was shown, as well as the impact of endogenous maternal factors (*MTHFR* C677T polymorphism) and exogenous maternal factors (body mass index and dietary habits such as folate intake) on maternal LINE-1 methylation and on the occurrence of CHD in children with DS. Study showed that the *MTHFR* genotype/diet combination and BMI were significantly associated with LINE-1 methylation in mothers of children with DS/CHD^+ [5]. Recently, micro-RNA signatures discordant for CHD in monozygotic twins were observed [112].

4.3 DNA methylation in developmental exposure to the maternal environment and diet

It has been suggested that disease risk of long-term health outcomes may be in part determined by maternal (in utero effects of environmental exposures, toxins/nutrition) [21, 113] and paternal diet [114, 115].

Birth defects occur in 6–10% of babies born to mothers with pregestational diabetes, which is a significant health problem. It has been demonstrated that exposure to maternal diabetes during pregnancy changes gene expression levels in the mouse embryo, disrupting essential cellular activities [116], and could lead to disruption of crucial epithelial and mesenchymal cell interactions in developing kidney, leading to kidney and urinary tract malformation [117]. Underlying mechanisms are still unknown. There is a proposed lack of precision in the developmental program, which is essential for organogenesis induced by hyperglycemia effects on oxidative stress. That exposure to a diabetic intrauterine environment during pregnancy could be teratogenic by leading to defects like CAKUT in the fetus and associate with metabolic or cardiovascular diseases in later life [118–121].

Changes in maternal dietary FA can affect the DNA methylation patterns of offspring in mice [61]. The *agouti* mouse is a best-studied example [122]. Recently, in the human genome, loci were found to show differential methylation in response to season of birth that is similar to the *agouti* locus, but the identity of the causative agent for the changes in DNA methylation is unclear [123]. Recent study examined the prospective association between multivitamin supplementation during pregnancy and maternal plasma folate/vitamin B_{12} levels at birth and child's autism spectrum disorder (ASD) risk. Moderate (3–5 times/week) self-reported supplementation during pregnancy was associated with decreased risk of ASD, consistent with previous findings. But, extremely high maternal plasma folate and B_{12} levels at birth were associated with ASD risk. This study raises new questions about the impact of extremely elevated levels of plasma folate and B_{12} exposure in utero on early brain development [124]. However, study on postmortem cortical brain samples reveals that global DNA methylation was markedly enriched in ASD brains [125]. In some diseases, methylation mosaicism was found to be present. This is a common phenomenon in Fragile X syndrome (FXS). A decreased gene expression was found to be a main contributor to the cognitive impairment observed in the study of 12 FXS males with atypical mosaicism, seven of whom presented with ASD [126].

5. Epigenetic pattern transmission from parent to offspring: understanding disease inheritance

The heritability of epigenetic modifications, including histone modifications and DNA methylation, provides a memory of cell function and identity. Transmission of epigenetic information to subsequent generations may provide evolutionary mechanisms that impact on adaptation to changed environment. Defining the mechanisms that establish and regulate the transmission of epigenetic information from parent to offspring is critical for understanding disease heredity. Detection of modified methylation patterns is important in inappropriate imprinting of certain either maternal or paternal genes, which are "turned on" by epigenetic phenomenon that leads to diseases such as Angelman syndrome and Prader-Willi syndrome. Methylation patterns with detrimental effects on development have been established for disorders of methylation, by several groups of researchers [127, 128]. One of the developed blood tests (EpiSign) claims to diagnose 19 congenital diseases [129]. Also, it is important to establish the potential for epigenomic drugs that have an impact on the germline epigenome and subsequent offspring [130, 131]. Currently, the molecular pathways that regulate epigenetic information in the germline and its transmission to offspring are poorly understood. Recent study reveals a novel role for the histone-modifying complex, PRC2, in maternal intergenerational transmission of epigenetic effects on offspring, with important implications for understanding disease inheritance [115]. PRC2 is involved in the regulation of many fundamental biological processes and is especially essential for embryonic stem cells. However, how the formation and function of PRC2 are regulated is mostly unknown. Recent findings identify miR-323-3p as a new regulator for PRC2, providing a new approach for regulating PRC2 activity via microRNAs [132]. Specific epigenetic pattern was observed to be essential in the development of CHD and CAKUT. Impaired transcriptional profiles in individuals with CHDs [133] and CAKUT [134, 135] were shown to be affected by epigenetic regulators of gene expression, using bioinformatical analysis and integrated prediction algorithms [136]. The miRNA-145 expression was confirmed in infants with CHD that negatively regulates gene expression important for heart development [133]. The altered hsa-miR-144 expression was, for the first time, identified in CAKUT and could be connected with biological processes crucial for normal development of kidney and urinary tract [135]. Although the importance of mothers' health prior to conception and during pregnancy is now well accepted, recent data also implicate fathers' health/nutritional status (overnutrition, undernutrition, and other forms of stress) in contribution to the risk of metabolic disease and obesity in offspring. Epigenetic paternal inheritance of chronic disease provides novel opportunities for multigenerational disease prevention [137]. Germ cell-dependent mechanisms have recently been linked to these intergenerational effects. There is increasing evidence that disruptions in male germ cell epigenetic reprogramming are associated with offspring abnormalities. Adequate supply of methyl donors is required in the fetal period, which is the critical time of DNA methylation pattern acquisition for developing male germ cells. In addition, DNA methylation patterns continue to be remodeled postnatal during spermatogenesis. Previous studies have shown that lifetime (prenatal and postnatal) folic acid deficiency and high-dose supplementation can alter the DNA methylation in sperm [138]. Recent study examined the genome-wide DNA methylation patterns in placentas and embryos in correlation with maternal FA supplementation in the prevention of CAs associated with assisted reproductive technologies (ART). Results demonstrate dose-dependent and sex-specific effects of FA intake; moderate dose of FA supplements may be optimal in ART for both sexes [139]. Even more, recent data suggest that genome-wide DNA methylation in the placentas from preterm infants could be associated with maternal socioeconomic status [140]. On the other hand, genomic information was identical in monozygotic twins, but they could be discordant for congenital renal agenesis which could be a consequence of epigenomic regulation of gene expression [141].

6. Future perspectives

CAs are complex traits with polygenic, epigenetic, and environmental components. Advances in human DNA methylation research and growing epigenetic data offer a new avenue for the translation of research to clinical applications. Current methylome analysis has been helpful in major human diseases revealing an epigenetic influence, but current approaches are inadequate for the translation of these advances to clinical diagnostics. There is a need to deal with big data in modern genomic medicine, so bioinformatics and applied mathematics are of a fundamental help in simulation studies and tests of methylome datasets. Signal detection theory and machine learning approaches applied on methylome datasets from ASD patients demonstrate high discriminatory power for the methylation signal induced by disease [142]. Even more, advanced machine learning analysis includes a combination of active learning and imbalanced class learning and deep learning to develop a more efficient feature selection process and for the generation and simultaneous computation of any genomic or biological dataset applied to medicine [143]. This approach demonstrates the feasibility in clinical diagnostics. Genetic risk scores (GRS) are widely used for risk prediction in complex diseases. Evidence is growing that methylation risk scores (MRS) may be constructed for multiple health purposes. MRS is defined as weighted sums of the individual's methylation markers' beta values of a preselected number of CpG sites and can be useful in interaction and mediation analyses, for environmental exposures as biomarker, and for prediction of individual risks of disease predisposition or treatment success [144]. As we know that methylation data is specific (for different tissues) and sensitive to confounding factor, e.g., by age or sex, adaption of current GRS approaches is complex and needs deep profiling in construction of such risk scores. The analysis of whole biomarker genomic and epigenomic regions and prediction of disease predisposition, course and therapy response by risk scores could in future suffice for a diagnostic and decreasing cost of patients' treatment.

7. Conclusion

The heritability of epigenetic modifications, including histone modifications and DNA methylation, provides a memory of cell function and identity. The dynamic reprogramming and other epigenetic patterns which could affect normal patterns of gene expression/genome stability during development could lead to an increased risk of CAs or complex diseases later in life. The sperm- and egg-specific DMT established during gametogenesis at imprint and non-imprint loci are susceptible to environmental factors. Embryogenesis may also present a window of opportunity for environmental influences on DNA methylation pattern. Changes in maternal dietary FA can affect the DNA methylation of offspring that could affect CA development. LINE-1 hypomethylation in the parental germline might also significantly affect genome stability during the fetal development. The *MTHFR* T carriers have lower enzyme activity, and dose of supplementary FA shows a positive relationship with DNA methylation patterns in humans. The lowest values of LINE-1 methylation, the surrogate marker for global DNA methylation, were observed in mothers with MTHFR 677 CT+TT genotype and low dietary folate, suggesting the possibility of impaired DNA methylation causing maternally derived trisomy 21. Also, *MTHFR* genotype/diet and BMI combination influence LINE-1 methylation in mothers that could be a risk factor for DS/CHD^{+} development in children. The studies discussed in this chapter provide new evidence to support nutrigenomic personalized FA treatment of mothers with risk genotype to prevent global DNA hypomethylation as potential underlying mechanism of CA development.

Acknowledgements

The work was funded by Grant No. OI175085, Ministry of Education, Science and Technological Development, Republic of Serbia.

Author details

Aleksandra Stanković
Laboratory for Radiobiology and Molecular Genetics, "Vinca" Institute of Nuclear Sciences, University of Belgrade, Belgrade, Serbia

*Address all correspondence to: alexas@vinca.rs

References

[1] WHO. Available from: https://www.who.int/news-room/fact-sheets/detail/congenital-anomalies

[2] Morris JK, Springett AL, Greenlees R, Loane M, Addor MC, Arriola L, et al. Trends in congenital anomalies in Europe from 1980 to 2012. PLoS ONE. 2018;**13**(4):e0194986. DOI: 10.1371/journal.pone.0194986

[3] Chowdhury S, Erickson SW, MacLeod SL, Cleves MA, Hu P, Karim MA, et al. Maternal genome-wide DNA methylation patterns and congenital heart defects. PLoS ONE. 2011;**6**(1):e16506. DOI: 10.1371/journal.pone.0016506

[4] Chowdhury S, Cleves MA, MacLeod SL, James SJ, Zhao W, Hobbs CA. Maternal DNA hypomethylation and congenital heart defects. Birth Defects Research. Part A, Clinical and Molecular Teratology. 2011;**91**:69-76. DOI: 10.1002/bdra.20761

[5] Babić Božović I, Stanković A, Živković M, Vraneković J, Mahulja-Stamenković V, et al. Maternal LINE-1 DNA methylation and congenital heart defects in down syndrome. Frontiers in Genetics. 2019;**10**:41. DOI: 10.3389/fgene.2019.00041

[6] Božović IB, Stanković A, Živković M, Vraneković J, Kapović M, Brajenović-Milić B. Altered LINE-1 methylation in mothers of children with Down syndrome. PLoS ONE. 2015;**10**(5):e0127423. DOI: 10.1371/journal.pone.0127423

[7] Finnell RH, Blom HJ, Shaw GM. Does global hypomethylation contribute to susceptibility to neural tube defects? The American Journal of Clinical Nutrition. 2010;**91**(5):1153-1154. DOI: 10.3945/ajcn.2010.29534

[8] Blom HJ, Shaw GM, den Heijer M, Finnell RH. Neural tube defects and folate: Case far from closed. Nature Reviews. Neuroscience. 2006;7(9):724-731. DOI: 10.1038/nrn1986

[9] Patel SR, Kim D, Levitan I, Dressler GR. The BRCT-domain containing protein PTIP links PAX2 to a histone H3, lysine 4 methyltransferase complex. Developmental Cell. 2007;**13**(4):580-592. DOI: 10.1016/j.devcel.2007.09.004

[10] Chen S, Bellew C, Yao X, Stefkova J, Dipp S, Saifudeen Z, et al. Histone deacetylase (HDAC) activity is critical for embryonic kidney gene expression, growth, and differentiation. The Journal of Biological Chemistry. 2011;**286**(37):32775-32789. DOI: 10.1074/jbc.M111.248278

[11] Adams DJ, Clark DA. Common genetic and epigenetic syndromes. Pediatric Clinics of North America. 2015;**62**(2):411-426. DOI: 10.1016/j.pcl.2014.11.005

[12] Skinner MK. Environmental epigenetic transgenerational inheritance and somatic epigenetic mitotic stability. Epigenetics. 2011;**6**(7):838-842. DOI: 10.4161/epi.6.7.16537

[13] van der Linde D, Konings EE, Slager MA, Witsenburg M, Helbing WA, Takkenberg JJ, et al. Birth prevalence of congenital heart disease worldwide: A systematic review and meta-analysis. Journal of the American College of Cardiology. 2011;**58**(21):2241-2247. DOI: 10.1016/j.jacc.2011.08.025

[14] Dolk H, Loane M, Garne E, European Surveillance of Congenital Anomalies (EUROCAT) Working Group. Congenital heart defects in Europe: Prevalence and perinatal mortality, 2000 to 2005. Circulation.

2011;**123**(8):841-849. DOI: 10.1161/CIRCULATIONAHA.110.958405

[15] Morris JK, Garne E, Wellesley D, Addor MC, Arriola L, Barisic I, et al. Major congenital anomalies in babies born with Down syndrome: A EUROCAT population-based registry study. American Journal of Medical Genetics. Part A. 2014;**164A**(12):2979-2986. DOI: 10.1002/ajmg.a.36780

[16] Freeman SB, Bean LH, Allen EG, Tinker SW, Locke AE, Druschel C, et al. Ethnicity, sex, and the incidence of congenital heart defects: A report from the National Down Syndrome Project. Genetics in Medicine. 2008;**10**(3):173-180. DOI: 10.1097/GIM.0b013e3181634867

[17] Pierpont ME, Brueckner M, Chung WK, Garg V, Lacro RV, McGuire AL, et al. Genetic basis for congenital heart disease: Revisited: A scientific statement from the American Heart Association. Circulation. 2018;**138**(21):e653-e711. DOI: 10.1161/CIR.0000000000000606

[18] Levin BL, Varga E. MTHFR: Addressing genetic counseling dilemmas using evidence-based literature. Journal of Genetic Counseling. 2016;**25**(5):901-911. DOI: 10.1007/s10897-016-9956-7

[19] Flom JD, Ferris JS, Liao Y, Tehranifar P, Richards CB, Cho YH, et al. Prenatal smoke exposure and genomic DNA methylation in a multiethnic birth cohort. Cancer Epidemiology, Biomarkers & Prevention. 2011;**20**(12):2518-2523. DOI: 10.1158/1055-9965.EPI-11-0553

[20] Terry MB, Delgado-Cruzata L, Vin-Raviv N, Wu HC, Santella RM. DNA methylation in white blood cells: Association with risk factors in epidemiologic studies. Epigenetics. 2011;**6**(7):828-837. DOI: 10.4161/epi.6.7.16500

[21] Dimasuay KG, Boeuf P, Powell TL, Jansson T. Placental responses to changes in the maternal environment determine fetal growth. Frontiers in Physiology. 2016;7:12. DOI: 10.3389/fphys.2016.00012

[22] Argyraki M, Damdimopoulou P, Chatzimeletiou K, Grimbizis GF, Tarlatzis BC, Syrrou M, et al. In-utero stress and mode of conception: Impact on regulation of imprinted genes, fetal development and future health. Human Reproduction Update. 2019;**25**(6):777-801. DOI: 10.1093/humupd/dmz025

[23] Pogribny IP, Beland FA. DNA hypomethylation in the origin and pathogenesis of human diseases. Cellular and Molecular Life Sciences. 2009;**66**(14):2249-2261. DOI: 10.1007/s00018-009-0015-5

[24] Robertson KD. DNA methylation and human disease. Nature Reviews. Genetics. 2005;**6**(8):597-610. DOI: 10.1038/nrg1655

[25] Matsha TE, Pheiffer C, Mutize T, Erasmus RT, Kengne AP. Glucose tolerance, MTHFR C677T and NOS3 G894T polymorphisms, and global DNA methylation in mixed ancestry African individuals. Journal Diabetes Research. 2016;**2016**:8738072. DOI: 10.1155/2016/8738072

[26] Song J, Rechkoblit O, Bestor TH, Patel DJ. Structure of DNMT1-DNA complex reveals a role for autoinhibition in maintenance DNA methylation. Science. 2011;**331**(6020):1036-1040. DOI: 10.1126/science.1195380

[27] Edwards JR, O'Donnell AH, Rollins RA, Peckham HE, Lee C, Milekic MH, et al. Chromatin and sequence features that define the fine and gross structure of genomic methylation patterns. Genome Research. 2010;**20**(7):972-980. DOI: 10.1101/gr.101535.109

[28] Bestor TH, Edwards JR, Boulard M. Notes on the role of dynamic DNA methylation in mammalian development. Proceedings of the National Academy of Sciences of the United States of America. 2015;**112**:6796-6799

[29] Portela A, Esteller M. Epigenetic modifications and human disease. Nature Biotechnology. 2010;**28**(10):1057-1068. DOI: 10.1038/nbt.1685

[30] Yarychkivska O, Shahabuddin Z, Comfort N, Boulard M, Bestor TH. BAH domains and a histone-like motif in DNA methyltransferase 1 (DNMT1) regulate *de novo* and maintenance methylation *in vivo*. The Journal of Biological Chemistry. 2018;**293**(50):19466-19475. DOI: 10.1074/jbc.RA118.004612

[31] Tahiliani M, Koh KP, Shen Y, Pastor WA, Bandukwala H, Brudno Y, et al. Conversion of 5-methylcytosine to 5-hydroxymethylcytosine in mammalian DNA by MLL partner TET1. Science. 2009;**324**:930-935. DOI: 10.1126/science.1170116

[32] Edwards JR, Yarychkivska O, Boulard M, Bestor TH. DNA methylation and DNA methyltransferases. Epigenetics & Chromatin. 2017;**10**:23. DOI: 10.1186/s13072-017-0130-8

[33] Krzyzewska IM, Maas SM, Henneman P, Lip KVD, Venema A, Baranano K, et al. A genome-wide DNA methylation signature for SETD1B-related syndrome. Clinical Epigenetics. 2019;**11**(1):156. DOI: 10.1186/s13148-019-0749-3

[34] Krzyzewska IM, Alders M, Maas SM, Bliek J, Venema A, Henneman P, et al. Genome-wide methylation profiling of Beckwith-Wiedemann syndrome patients without molecular confirmation after routine diagnostics. Clinical Epigenetics. 2019;**11**(1):53. DOI: 10.1186/s13148-019-0649-6

[35] Docherty LE, Rezwan FI, Poole RL, Jagoe H, Lake H, Lockett GA, et al. Genome-wide DNA methylation analysis of patients with imprinting disorders identifies differentially methylated regions associated with novel candidate imprinted genes. Journal of Medical Genetics. 2014;**51**(4):229-238. DOI: 10.1136/jmedgenet-2013-102116

[36] Gregg C, Zhang J, Butler JE, Haig D, Dulac C. Sex-specific parent-of-origin allelic expression in the mouse brain. Science. 2010;**329**:682-685. DOI: 10.1126/science.1190831

[37] Guilleret I, Yan P, Grange F, Braunschweig R, Bosman FT, Benhattar J. Hypermethylation of the human telomerase catalytic subunit (hTERT) gene correlates with telomerase activity. International Journal of Cancer. 2002;**101**:335-341. DOI: 10.1002/ijc.10593

[38] Axelrad MD, Budagov T, Atzmon G. Telomere length and telomerase activity: A Yin and Yang of cell senescence. Journal of Visualized Experiments. 2013;**75**:e50246. DOI: 10.3791/50246

[39] Joksic I, Vujic D, Guc-Scekic M, Leskovac A, Petrovic S, Ojani M, et al. Dysfunctional telomeres in primary cells from Fanconi anemia FANCD2 patients. Genome Integrity. 2012;**3**:6. DOI: 10.1186/2041-9414-3-6

[40] Wilson AS, Power BE, Molloy PL. DNA hypomethylation and human diseases. Biochimica et Biophysica Acta. 2007;**1775**:138-162. DOI: 10.1016/j.bbcan.2006.08.007

[41] Herrera LA, Prada D, Andonegui MA, Dueñas-González A. The epigenetic origin of aneuploidy. Current Genomics. 2008;**9**:43-50. DOI: 10.2174/138920208783884883

[42] Lander ES, Linton LM, Birren B, Nusbaum C, Zody MC, Baldwin J, et al.

Initial sequencing and analysis of the human genome. Nature. 2001;**409**:860-921. DOI: 10.1038/35057062

[43] Kazazian HH Jr. Mobile elements: Drivers of genome evolution. Science. 2004;**303**:1626-1632. DOI: 10.1126/science.1089670

[44] Suzuki MM, Bird A. DNA methylation landscapes: Provocative insights from epigenomics. Nature Reviews. Genetics. 2008;**9**:465-476. DOI: 10.1038/nrg2341

[45] Weisenberger DJ, Campan M, Long TI, Kim M, Woods C, Fiala E, et al. Analysis of repetitive element DNA methylation by MethyLight. Nucleic Acids Research. 2005;**33**:6823-6836. DOI: 10.1093/nar/gki987

[46] Yang AS, Estécio MR, Doshi K, Kondo Y, Tajara EH, Issa JP. A simple method for estimating global DNA methylation using bisulfite PCR of repetitive DNA elements. Nucleic Acids Research. 2004;**32**:e38. DOI: 10.1093/nar/gnh032

[47] Chen JM, Stenson PD, Cooper DN, Ferec C. A systematic analysis of LINE-1 endonuclease-dependent retrotranspositional events causing human genetic disease. Human Genetics. 2005;**117**:411-427. DOI: 10.1007/s00439-005-1321-0

[48] Woo HD, Kim J. Global DNA hypomethylation in peripheral blood leukocytes as a biomarker for cancer risk: A meta-analysis. PLoS ONE. 2012;7:e34615. DOI: 10.1371/journal.pone.0034615

[49] Barchitta M, Quattrocchi A, Maugeri A, Vinciguerra M, Agodi A. LINE-1 hypomethylation in blood and tissue samples as an epigenetic marker for cancer risk: A systematic review and meta-analysis. PLoS ONE. 2014;**9**:e109478. DOI: 10.1371/journal.pone.0109478

[50] Baccarelli A, Rienstra M, Benjamin EJ. Cardiovascular epigenetics: Basic concepts and results from animal and human studies. Circulation. Cardiovascular Genetics. 2010;**3**:567-573. DOI: 10.1161/CIRCGENETICS.110.958744

[51] Agodi A, Barchitta M, Maugeri A, Basile G, Zamboni M, Bernardini G, et al. Unveiling the role of DNA methylation in kidney transplantation: Novel perspectives toward biomarker identification. BioMed Research International. 2019;**2019**:1602539. DOI: 10.1155/2019/1602539

[52] Filkowski JN, Ilnytskyy Y, Tamminga J, Koturbash I, Golubov A, Bagnyukova T, et al. Hypomethylation and genome instability in the germline of exposed parents and their progeny is associated with altered miRNA expression. Carcinogenesis. 2010;**31**:1110-1115. DOI: 10.1093/carcin/bgp300

[53] Barchitta M, Maugeri A, Quattrocchi A, Agrifoglio O, Agodi A. The role of miRNAs as biomarkers for pregnancy outcomes: A comprehensive review. International Journal of Genomics. 2017;**2017**:8067972. DOI: 10.1155/2017/8067972

[54] Hajkova P, Erhardt S, Lane N, Haaf T, El-Maarri O, Reik W, et al. Epigenetic reprogramming in mouse primordial germ cells. Mechanisms of Development. 2002;**117**:15-23. DOI: 10.1016/s0925-4773(02)00181-8

[55] Szabó PE, Hübner K, Schüler H, Mann JR. Allele-specific expression of imprinted genes in mouse migratory primordial germ cells. Mechanisms of Development. 2002;**115**:157-160. DOI: 10.1016/s0925-4773(02)00087-4

[56] Maatouk DM, Kellam LD, Mann MR, Lei H, Li E, Bartolomei MS, et al. DNA methylation is a primary

mechanism for silencing postmigratory primordial germ cell genes in both germ cell and somatic cell lineages. Development. 2006;**133**:3411-3418. DOI: 10.1242/dev.02500

[57] Lees-Murdock DJ, De Felici M, Walsh CP. Methylation dynamics of repetitive DNA elements in the mouse germ cell lineage. Genomics. 2003;**82**:230-237. DOI: 10.1016/s0888-7543(03)00105-8

[58] Sasaki H, Matsui Y. Epigenetic events in mammalian germ-cell development: Reprogramming and beyond. Nature Reviews. Genetics. 2008;**9**:129-140. DOI: 10.1038/nrg2295

[59] Feng S, Jacobsen SE, Reik W. Epigenetic reprogramming in plant and animal development. Science. 2010;**330**:622-627. DOI: 10.1126/science.1190614

[60] Ji H, Ehrlich LI, Seita J, Murakami P, Doi A, Lindau P, et al. Comprehensive methylome map of lineage commitment from haematopoietic progenitors. Nature. 2010;**467**:338-342. DOI: 10.1038/nature09367

[61] Hirasawa R, Sasaki H. Dynamic transition of Dnmt3b expression in mouse pre- and early post-implantation embryos. Gene Expression Patterns. 2009;**9**(1):27-30. DOI: 10.1016/j.gep.2008.09.002

[62] Waterland RA, Kellermayer R, Rached MT, Tatevian N, Gomes MV, Zhang J, et al. Epigenomic profiling indicates a role for DNA methylation in early postnatal liver development. Human Molecular Genetics. 2009;**18**:3026-3038. DOI: 10.1093/hmg/ddp241

[63] Fulka H, Mrazek M, Tepla O, Fulka J Jr. DNA methylation pattern in human zygotes and developing embryos. Reproduction. 2004;**128**:703-708. DOI: 10.1530/rep.1.00217

[64] Petrussa L, Van de Velde H, De Rycke M. Dynamic regulation of DNA methyltransferases in human oocytes and preimplantation embryos after assisted reproductive technologies. Molecular Human Reproduction. 2014;**20**(9):861-874. DOI: 10.1093/molehr/gau049

[65] Dolinoy DC, Das R, Weidman JR, Jirtle RL. Metastable epialleles, imprinting, and the fetal origins of adult diseases. Pediatric Research. 2007;**61**:30R-37R. DOI: 10.1203/pdr.0b013e31804575f7

[66] Li Y, Xie C, Murphy SK, Skaar D, Nye M, Vidal AC, et al. Lead exposure during early human development and DNA methylation of imprinted gene regulatory elements in adulthood. Environmental Health Perspectives. 2016;**124**(5):666-673. DOI: 10.1289/ehp.1408577

[67] Zhang W, Li M, Sun F, Xu X, Zhang Z, Liu J, et al. Association of sperm methylation at *LINE-1*, four candidate genes, and nicotine/alcohol exposure with the risk of infertility. Frontiers in Genetics. 2019;**10**:1001. DOI: 10.3389/fgene.2019.01001

[68] Yang X, Lay F, Han H, Jones PA. Targeting DNA methylation for epigenetic therapy. Trends in Pharmacological Sciences. 2010;**31**:536-546

[69] Barchitta M, Maugeri A, Magnano San Lio R, Favara G, La Rosa MC, La Mastra C, et al. Dietary patterns are associated with leukocyte LINE-1 methylation in women: A cross-sectional study in Southern Italy. Nutrients. 2019;**11**:E1843. DOI: 10.3390/nu11081843

[70] Agodi A, Barchitta M, Quattrocchi A, Maugeri A, Canto C, Marchese AE, et al. Low fruit consumption and folate deficiency are associated with LINE-1

hypomethylation in women of a cancer-free population. Genes & Nutrition. 2015;**10**:480. DOI: 10.1007/s12263-015-0480-4

[71] Zhang FF, Morabia A, Carroll J, Gonzalez K, Fulda K, Kaur M, et al. Dietary patterns are associated with levels of global genomic DNA methylation in a cancer-free population. The Journal of Nutrition. 2011;**141**:1165-1171. DOI: 10.3945/jn.110.134536

[72] Saura-Calixto F, Goñi I. Definition of the Mediterranean diet based on bioactive compounds. Critical Reviews in Food Science and Nutrition. 2009;**49**:145-152. DOI: 10.1080/10408390701764732

[73] Liu RH. Dietary bioactive compounds and their health implications. Journal of Food Science. 2013;**78**:A18-A25. DOI: 10.1111/1750-3841.12101

[74] Steluti J, Palchetti CZ, Miranda AM, Fisberg RM, Marchioni DM. DNA methylation and one-carbon metabolism related nutrients and polymorphisms: Analysis after mandatory flour fortification with folic acid. The British Journal of Nutrition. 2020;**123**:23-29. DOI: 10.1017/S0007114519002526

[75] Berry RJ, Bailey L, Mulinare J, Bower C. Fortification of flour with folic acid. Food and Nutrition Bulletin. 2010;**31**:S22-S35. DOI: 10.1177/15648265100311S103

[76] Barchitta M, Maugeri A, Quattrocchi A, Barone G, Mazzoleni P, Catalfo A, et al. Mediterranean diet and particulate matter exposure are associated with LINE-1 methylation: Results from a cross-sectional study in women. Frontiers in Genetics. 2018;**9**:514. DOI: 10.3389/fgene.2018.00514

[77] de la Rocha C, Pérez-Mojica JE, León SZ, Cervantes-Paz B, Tristán-Flores FE, Rodríguez-Ríos D, et al. Associations between whole peripheral blood fatty acids and DNA methylation in humans. Scientific Reports. 2016;**6**:25867. DOI: 10.1038/srep25867

[78] Milenkovic D, VandenBerghe W, Boby C, Leroux C, Declerck K, Szarc vel Szic K, et al. Dietary flavanols modulate the transcription of genes associated with cardiovascular pathology without changes in their DNA methylation state. PLoS ONE. 2014;**9**(4):e95527. DOI: 10.1371/journal.pone.0095527

[79] Rampersaud GC, Kauwell GP, Hutson AD, Cerda JJ, Bailey LB. Genomic DNA methylation decreases in response to moderate folate depletion in elderly women. The American Journal of Clinical Nutrition. 2000;**72**:998-1003. DOI: 10.1093/ajcn/72.4.998

[80] Jacob RA, Gretz DM, Taylor PC, James SJ, Pogribny IP, Miller BJ, et al. Moderate folate depletion increases plasma homocysteine and decreases lymphocyte DNA methylation in postmenopausal women. The Journal of Nutrition. 1998;**128**:1204-1212. DOI: 10.1093/jn/128.7.1204

[81] Shelnutt KP, Kauwell GP, Gregory JF 3rd, Maneval DR, Quinlivan EP, Theriaque DW, et al. Methylenetetrahydrofolate reductase 677C/T polymorphism affects DNA methylation in response to controlled folate intake in young women. The Journal of Nutritional Biochemistry. 2004;**15**:554-560. DOI: 10.1016/j.jnutbio.2004.04.003

[82] Friso S, Choi SW, Girelli D, Mason JB, Dolnikowski GG, Bagley PJ, et al. A common mutation in the 5,10-methylenetetrahydrofolate reductase gene affects genomic DNA methylation through an interaction with folate status. Proceedings of the National Academy of Sciences of the United States of America.

2002;**99**:5606-5611. DOI: 10.1073/pnas.062066299

[83] Shin W, Yan J, Abratte CM, Vermeylen F, Caudill MA. Choline intake exceeding current dietary recommendations preserves markers of cellular methylation in a genetic subgroup of folate-compromised men. The Journal of Nutrition. 2010;**140**:975-980. DOI: 10.3945/jn.110.121186

[84] Hoffman DR, Marion DW, Cornatzer WE, Duerre JA. S-adenosylmethionine and S-adenosylhomocystein metabolism in isolated rat liver. Effects of L-methionine, L-homocystein, and adenosine. The Journal of Biological Chemistry. 1980;**255**:10822-10827

[85] James SJ, Melnyk S, Pogribna M, Pogribny IP, Caudill MA. Elevation in S-adenosylhomocysteine and DNA hypomethylation: Potential epigenetic mechanism for homocysteine-related pathology. The Journal of Nutrition. 2002;**132**:2361S-23616S. DOI: 10.1093/jn/132.8.2361S

[86] Yi P, Melnyk S, Pogribna M, Pogribny IP, Hine RJ, James SJ. Increase in plasma homocysteine associated with parallel increases in plasma S-adenosylhomocysteine and lymphocyte DNA hypomethylation. The Journal of Biological Chemistry. 2000;**102**:29318-29323. DOI: 10.1074/jbc.M002725200

[87] Stover PJ. One-carbon metabolism-genome interactions in folate associated pathologies. The Journal of Nutrition. 2009;**139**:2402-2405. DOI: 10.3945/jn.109.113670

[88] Miranda TB, Jones PA. DNA methylation: The nuts and bolts of repression. Journal of Cellular Physiology. 2007;**213**:384-390. DOI: 10.1002/jcp.21224

[89] Erickson JD. Folic acid and prevention of spina bifida and anencephaly. 10 years after the U.S. Public Health Service recommendation. MMWR—Recommendations and Reports. 2002;**51**:1-3

[90] Lamprecht SA, Lipkin M. Chemoprevention of colon cancer by calcium, vitamin D and folate: Molecular mechanisms. Nature Reviews. Cancer. 2003;**3**:601-614. DOI: 10.1038/nrc1144

[91] Holmquist C, Larsson S, Wolk A, de Faire U. Multivitamin supplements are inversely associated with risk of myocardial infarction in men and women-Stockholm Heart Epidemiology Program (SHEEP). The Journal of Nutrition. 2003;**133**:2650-2654. DOI: 10.1093/jn/133.8.2650

[92] Lacasaña M, Blanco-Muñoz J, Borja-Aburto VH, Aguilar-Garduño C, Rodríguez-Barranco M, Sierra-Ramirez JA, et al. Effect on risk of anencephaly of gene-nutrient interactions between methylenetetrahydrofolate reductase C677T polymorphism and maternal folate, vitamin B_{12} and homocysteine profile. Public Health Nutrition. 2012;**15**:1419-1428. DOI: 10.1017/S136898001100334X

[93] ElGendy K, Malcomson FC, Lara JG, Bradburn DM, Mathers JC. Effects of dietary interventions on DNA methylation in adult humans: Systematic review and meta-analysis. The British Journal of Nutrition. 2018;**120**:961-976. DOI: 10.1017/S000711451800243X

[94] Huang X, Qin X, Yang W, Liu L, Jiang C, Zhang X, et al. MTHFR gene and serum folate interaction on serum homocysteine lowering: Prospect for precision folic acid treatment. Arteriosclerosis, Thrombosis, and Vascular Biology. 2018;**38**:679-685. DOI: 10.1161/ATVBAHA.117.310211

[95] Caffrey A, McNulty H, Irwin RE, Walsh CP, Pentieva K. Maternal folate nutrition and offspring health: Evidence and current controversies. The Proceedings of the Nutrition Society. 2018;**26**:1-13. DOI: 10.1017/S0029665118002689

[96] Chikakuda AT, Shin D, Comstock SS, Song S, Song WO. Compliance to prenatal iron and folic acid supplement use in relation to low birth weight in Lilongwe, Malawi. Nutrients. 2018;**10**:10. DOI: 10.3390/nu10091275

[97] Li Z, Mei Z, Zhang L, Li H, Zhang Y, Li N, et al. Effects of prenatal micronutrient supplementation on spontaneous preterm birth: A double-blind randomized controlled trial in China. American Journal of Epidemiology. 2017;**186**:318-325. DOI: 10.1093/aje/kwx094

[98] Obeid R, Oexle K, Rißmann A, Pietrzik K, Koletzko B. Folate status and health: Challenges and opportunities. Journal of Perinatal Medicine. 2016;**44**:261-268. DOI: 10.1515/jpm-2014-0346

[99] Kimura M, Umegaki K, Higuchi M, Thomas P, Fenech M. Methylenetetrahydrofolate reductase C677T polymorphism, folic acid and riboflavin are important determinants of genome stability in cultured human lymphocytes. The Journal of Nutrition. 2004;**134**:48-56. DOI: 10.1093/jn/134.1.48

[100] Liew SC, Gupta ED. Methylenetetrahydrofolate reductase (MTHFR) C677T polymorphism: Epidemiology, metabolism and the associated diseases. European Journal of Medical Genetics. 2015;**58**:1-10. DOI: 10.1016/j.ejmg.2014.10.004

[101] Frosst P, Blom HJ, Milos R, Goyette P, Sheppard CA, Matthews RG, et al. A candidate genetic risk factor for vascular disease: A common mutation in methylenetetrahydrofolate reductase. Nature Genetics. 1995;**10**(1):111-113. DOI: 10.1038/ng0595-111

[102] Sundstrom J, Sullivan L, D'Agostino RB, Jacques PF, Selhub J, Rosenberg IH, et al. Plasma homocysteine, hypertension incidence, and blood pressure tracking: The Framingham heart study. Hypertension. 2003;**42**:1100-1105. DOI: 10.1161/01.HYP.0000101690.58391.13

[103] Tsang BL, Devine OJ, Cordero AM, et al. Assessing the association between the methylenetetrahydrofolate reductase (MTHFR) 677C>T polymorphism and blood folate concentrations: A systematic review and meta-analysis of trials and observational studies. The American Journal of Clinical Nutrition. 2015;**101**:1286-1294

[104] Herrmann W, Obeid R, Schorr H, Zarzour W, Geisel J. Homocysteine, methylenetetrahydrofolate reductase C677T polymorphism and the B-vitamins: A facet of nature-nurture interplay. Clinical Chemistry and Laboratory Medicine. 2003;**41**:547-553. DOI: 10.1515/CCLM.2003.08

[105] Kaur A, Kaur A. Maternal MTHFR polymorphism (677 C-T) and risk of Down's syndrome child: Meta-analysis. Journal of Genetics. 2016;**95**:505-513. DOI: 10.1007/s12041-016-0657-7

[106] Božović IB, Vraneković J, Cizmarević NS, Mahulja-Stamenković V, Prpić I, Brajenović-Milić B. MTHFR C677T and A1298C polymorphisms as a risk factor for congenital heart defects in Down syndrome. Pediatrics International. 2011;**53**:546-550. DOI: 10.1111/j.1442-200X.2010.03310.x

[107] Vraneković J, Babić Bozović I, StarcevićCizmarević N, Buretić-Tomljanović A, Ristić S, Petrović O, et al. Functional inference of methylenetetrahydrofolate reductase

gene polymorphisms on enzyme stability as a potential risk factor for Down syndrome in Croatia. Disease Markers. 2010;**28**:293-298. DOI: 10.3233/DMA-2010-0704

[108] Coppedè F, Denaro M, Tannorella P, Migliore L. Increased MTHFR promoter methylation in mothers of Down syndrome individuals. Mutation Research. 2016;**787**:1-6. DOI: 10.1016/j.mrfmmm.2016.02.008

[109] Asim A, Agarwal S, Panigrahi I, Saiyed N, Bakshi S. MTHFR promoter hypermethylation may lead to congenital heart defects in Down syndrome. Intractable & Rare Diseases Research. 2017;**6**:295-298. DOI: 10.5582/irdr.2017.01068

[110] Axume J, Smith SS, Pogribny IP, Moriarty DJ, Caudill MA. The methylenetetrahydrofolate reductase 677TT genotype and folate intake interact to lower global leukocyte DNA methylation in young Mexican American women. Nutrition Research. 2007;**27**:13-17. DOI: 10.1016/j.nutres.2006.12.006

[111] Mishra J, Talwar S, Kaur L, Chandiok K, Yadav S, Puri M, et al. Differential global and MTHFR gene specific methylation patterns in preeclampsia and recurrent miscarriages: A case-control study from North India. Gene. 2019;**704**:68-73. DOI: 10.1016/j.gene.2019.04.036

[112] Abu-Halima M, Weidinger J, Poryo M, Henn D, Keller A, Meese E, et al. Micro-RNA signatures in monozygotic twins discordant for congenital heart defects. PLoS ONE. 2019;**14**(12):e0226164. DOI: 10.1371/journal.pone.0226164

[113] Gabbianelli R, Damiani E. Epigenetics and neurodegeneration: Role of early-life nutrition. The Journal of Nutritional Biochemistry. 2018;**57**:1-13. DOI: 10.1016/j.jnutbio.2018.01.014

[114] Gluckman PD, Hanson MA. Living with the past: Evolution, development, and patterns of disease. Science. 2004;**305**:1733-1736. DOI: 10.1126/science.1095292

[115] Stringer JM, Forster SC, Qu Z, Prokopuk L, O'Bryan MK, Gardner DK, et al. Reduced PRC2 function alters male germline epigenetic programming and paternal inheritance. Biochemistry and Molecular Biology. 2018;**16**:104. DOI: 10.1186/s12915-018-0569-5

[116] Pavlinkova G, Salbaum JM, Kappen C. Maternal diabetes alters transcriptional programs in the developing embryo. BMC Genomics. 2009;**10**:274. DOI: 10.1186/1471-2164-10-274

[117] Kanwar YS, Nayak B, Lin S, et al. Hyperglycemia: Its imminent effects on mammalian nephrogenesis. Pediatric Nephrology. 2005;**20**:858-866. DOI: 10.1007/s00467-005-1888-7

[118] Salbaum JM, Kappen C. Diabetic embryopathy: A role for the epigenome? Birth Defects Research. Part A, Clinical and Molecular Teratology. 2011;**91**:770-780. DOI: 10.1002/bdra.20807

[119] Feil R, Fraga MF. Epigenetics and the environment: Emerging patterns and implications. Nature Reviews. Genetics. 2012;**13**:97-109. DOI: 10.1038/nrg3142

[120] Vrachnis N, Antonakopoulos N, Iliodromiti Z, Dafopoulos K, Siristatidis C, Pappa KI, et al. Impact of maternal diabetes on epigenetic modifications leading to diseases in the offspring. Experimental Diabetes Research. 2012;**2012**:538474. DOI: 10.1155/2012/538474

[121] Franzago M, Fraticelli F, Stuppia L, Vitacolonna E. Nutrigenetics, epigenetics and gestational diabetes: Consequences in mother and child. Epigenetics. 2019;**14**(3):215-235. DOI: 10.1080/15592294.2019.1582277

[122] Waterland RA. Do maternal methyl supplements in mice affect DNA methylation of offspring? The Journal of Nutrition. 2003;**133**:238. DOI: 10.1093/jn/133.1.238

[123] Waterland RA, Kellermayer R, Laritsky E, Rayco-Solon P, Harris RA, Travisano M, et al. Season of conception in rural Gambia affects DNA methylation at putative human metastable epialleles. PLoS Genetics. 2010;**6**:e1001252. DOI: 10.1371/journal.pgen.1001252

[124] Raghavan R, Riley AW, Volk H, Caruso D, Hironaka L, Sices L, et al. Maternal multivitamin intake, plasma folate and vitamin B_{12} levels and autism spectrum disorder risk in offspring. Paediatric and Perinatal Epidemiology. 2018;**32**:100-111. DOI: 10.1111/ppe.12414

[125] Ellis SE, Gupta S, Moes A, West AB, Arking DE. Exaggerated CpH methylation in the autism-affected brain. Molecular Autism. 2017;**8**:6. DOI: 10.1186/s13229-017-0119-y

[126] Jiraanont P, Kumar M, Tang HT, Espinal G, Hagerman PJ, Hagerman RJ, et al. Size and methylation mosaicism in males with Fragile X syndrome. Expert Review of Molecular Diagnostics. 2017;**17**:1023-1032. DOI: 10.1080/14737159.2017.1377612

[127] Elhamamsy AR. Role of DNA methylation in imprinting disorders: An updated review. Journal of Assisted Reproduction and Genetics. 2017;**34**:549-562. DOI: 10.1007/s10815-017-0895-5

[128] Caliebe A, Richter J, Ammerpohl O, Kanber D, Beygo J, Bens S, et al. A familial disorder of altered DNA-methylation. Journal of Medical Genetics. 2014;**51**:407-412. DOI: 10.1136/jmedgenet-2013-102149

[129] EpiSign Complete South Carolina: Greenwood Genetic Center. 2019. Available from: https://www.ggc.org/test-finder-item/episign-complete)/

[130] Jarred EG, Bildsoe H, Western PS. Out of sight, out of mind? Germ cells and the potential impacts of epigenomic drugs. F1000Res. 2018;**7**:F1000 Faculty Rev-1967. DOI: 10.12688/f1000research.15935.1

[131] Prokopuk L, Western PS, StringerJM. Transgenerational epigenetic inheritance: Adaptation through the germline epigenome? Epigenomics. 2015;**7**:829-846. DOI: 10.2217/epi.15.36

[132] Zhang Y, Teng F, Luo GZ, Wang M, Tong M, Zhao X, et al. MicroRNA-323-3p regulates the activity of polycomb repressive complex 2 (PRC2) via targeting the mRNA of embryonic ectoderm development (Eed) gene in mouse embryonic stem cells. The Journal of Biological Chemistry. 2013;**288**:23659-23665. DOI: 10.1074/jbc.M113.475608

[133] Wang L, Tian D, Hu J, Xing H, Sun M, Wang J, et al. MiRNA-145 regulates the development of congenital heart disease through targeting FXN. Pediatric Cardiology. 2016;**37**:629-636. DOI: 10.1007/s00246-015-1325-z

[134] Jovanovic I, Zivkovic M, Kostic M, Krstic Z, Djuric T, Licastro D, et al. Transcriptome-driven integrative exploration of functional state of ureter tissue affected by CAKUT. Life Sciences. 2018;**212**:1-8. DOI: 10.1016/j.lfs.2018.09.042

[135] Jovanovic I, Zivkovic M, Kostic M, Krstic Z, Djuric T, Kolic I, et al. Transcriptome-wide based identification of miRs in congenital anomalies of the kidney and urinary tract (CAKUT) in children: The significant upregulation of tissue miR-144 expression. Journal of Translational

Medicine. 2016;**14**:193. DOI: 10.1186/s12967-016-0955-0

[136] Jovanović I, Zivković M, Jovanović J, Djurić T, Stanković A. The co-inertia approach in identification of specific microRNA in early and advanced atherosclerosis plaque. Medical Hypotheses. 2014;**83**:11-15. DOI: 10.1016/j.mehy.2014.04.019

[137] Isganaitis E, Suehiro H, Cardona C. Who's your daddy? Paternal inheritance of metabolic disease risk. Current Opinion in Endocrinology, Diabetes, and Obesity. 2017;**24**:47-55. DOI: 10.1097/MED.0000000000000307

[138] Ly L, Chan D, Aarabi M, Landry M, Behan NA, MacFarlane AJ, et al. Intergenerational impact of paternal lifetime exposures to both folic acid deficiency and supplementation on reproductive outcomes and imprinted gene methylation. Molecular Human Reproduction. 2017;**23**:461-477. DOI: 10.1093/molehr/gax029

[139] Rahimi S, Martel J, Karahan G, Angle C, Behan NA, et al. Moderate maternal folic acid supplementation ameliorates adverse embryonic and epigenetic outcomes associated with assisted reproduction in a mouse model. Human Reproduction. 2019;**34**:851-862. DOI: 10.1093/humrep/dez036

[140] Santos HP Jr, Bhattacharya A, Martin EM, Addo K, Psioda M, Smeester L, et al. Epigenome-wide DNA methylation in placentas from preterm infants: Association with maternal socioeconomic status. Epigenetics. 2019;**14**:751-765. DOI: 10.1080/15592294.2019.1614743

[141] Jin M, Zhu S, Hu P, Liu D, Li Q, Li Z, et al. Genomic and epigenomic analyses of monozygotic twins discordant for congenital renal agenesis. American Journal of Kidney Diseases. 2014;**64**:119-122. DOI: 10.1053/j.ajkd.2014.01.423

[142] Sanchez R, Yang X, Maher T, Mackenzie SA. Discrimination of DNA methylation signal from background variation for clinical diagnostics. International Journal of Molecular Sciences. 2019;**20**:5343. DOI: 10.3390/ijms20215343

[143] Holder LB, Haque MM, Skinner MK. Machine learning for epigenetics and future medical applications. Epigenetics. 2017;**12**:505-514. DOI: 10.1080/15592294.2017.1329068

[144] Hüls A, Czamara D. Methodological challenges in constructing DNA methylation risk scores. Epigenetics. 2020;**15**:1-11. DOI: 10.1080/15592294.2019.1644879

Chapter 4

Recent Insights into the Mechanisms of *De Novo* and Maintenance of DNA Methylation in Mammals

Motoko Unoki

Abstract

DNA methylation is one of the key epigenetic mechanisms essential for transcriptional regulation, silencing of transposable elements, and genome stabilization. Under physiological conditions, DNA methylation is erased and then established genome-wide during gametogenesis and embryogenesis. *De novo* DNA methylation by the enzymatic reaction of the *de novo* DNA methyltransferases (DNMTs), DNMT3A and DNMT3B, occurs during the establishment of DNA methylation patterns specific to each germ cell type or somatic cell type after the erasure. Once cell type-specific DNA methylation patterns are established during embryogenesis, which can extend to early childhood, the maintenance of DNA methyltransferase DNMT1 and its cofactor UHRF1 cooperatively maintain the pattern throughout the individual's lifetime. Recently, our group found that UHRF1 is also involved in *de novo* DNA methylation during oogenesis. Moreover, our group has identified two genes, *CDCA7* and *HELLS*, to be the causative genes of ICF syndrome, characterized by hypomethylation of centromeric and pericentromeric repetitive sequences. Because CDCA7/HELLS comprise a chromatin remodeling complex, there are evidently certain regions where chromatin remodeling is required to achieve maintenance of DNA methylation. In this chapter, the current situation with respect to our understanding of *de novo* and maintenance of DNA methylation mechanisms under physiological conditions in mammals is summarized.

Keywords: *de novo* DNA methylation, maintenance of DNA methylation, embryogenesis, gametogenesis, oocyte, PGC, ubiquitylation, UHRF1, DNMT1, DNMT3A, DNMT3B, DNMT3L, ICF syndrome, CDCA7, HELLS, chromatin remodeling

1. Introduction

Methylation at the C5 positions of cytosine (i.e., 5mC) in the CpG context (hereafter called DNA methylation) plays a major role in the transcriptional regulation of gene expression, the silencing of transposable elements (TEs), and genome integrity. The enzymatic activities catalyzing DNA methylation can be classified into two types. One is *de novo* DNA methylation, which is an activity by which methyl groups are added to cytosine at unmethylated DNA; *de novo* DNA methyltransferases DNMT3A and DNMT3B, together with their coactivator DNMT3L, are known to catalyze this reaction [1]. The other DNA methylation type is maintenance of DNA methylation, an activity by which unmethylated cytosine residues of hemi-methylated DNA (cytosine methylation on only one strand of the CG dyad) are methylated after DNA replication; DNA methyltransferase 1 (DNMT1) and its cofactor, ubiquitin-like, containing PHD and RING finger domains 1 (UHRF1), are responsible for this function [2]. In this chapter, the current status of knowledge of the two types of DNA methylation, including recent findings from our group, is summarized. In addition, possible mechanisms underlying the control of region-specific methylation by *de novo* DNMTs as well as possible maintenance of DNA methylation mechanisms, with or without chromatin remodeling, are discussed.

2. *De novo* DNA methylation

2.1 Timing of physiological *de novo* DNA methylation

De novo DNA methylation is a process by which methyl groups are added to unmethylated DNA at specific CpG sites, catalyzed by DNMT3A and DNMT3B [1]. In most differentiated cells, *de novo* DNA methylation is basically undesirable, since precise maintenance of the DNA methylation pattern, once established, is essential to sustain the appropriate functions of each cell type. Under physiological conditions, DNA methylation is widely erased, and then cell-type specific DNA methylation patterns are established during gametogenesis and embryogenesis [3] (**Figure 1**). During oogenesis and spermatogenesis, *de novo* DNA methylation occurs to establish oocyte- and sperm-specific DNA methylation patterns, following its elimination in primordial germ cells (PGCs). During post-implantation embryogenesis, which can extend to early childhood, *de novo* DNA methylation occurs to establish DNA methylation patterns specific to each cell type after they are erased in preimplantation embryos.

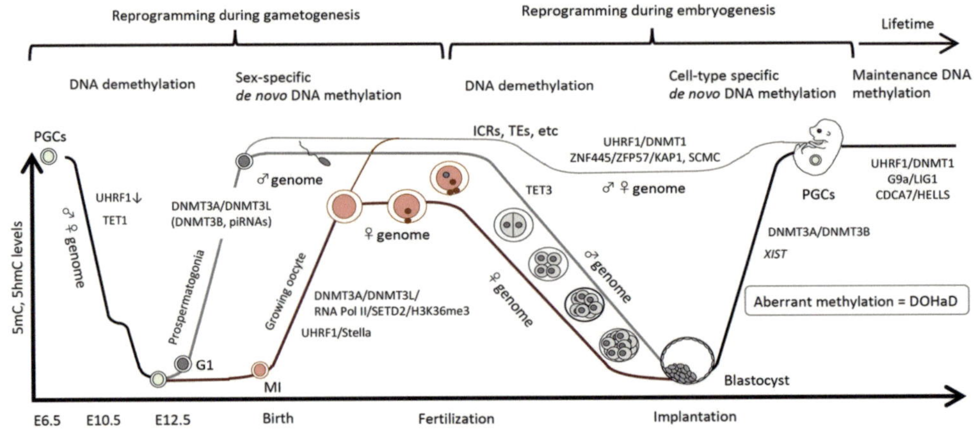

Figure 1.
Known factors involved in dynamic physiological demethylation and de novo DNA methylation during mouse gametogenesis and embryogenesis. DNA methylation is widely erased, and then sex-specific or cell type-specific DNA methylation patterns are established during gametogenesis and embryogenesis. Levels of DNA methylation of imprinting control regions (ICRs) and some transposable elements (TEs) are maintained in the face of genome-wide demethylation in preimplantation embryos. DOHaD, developmental origins of health and disease; SCMC, subcortical maternal complex; E6.5, embryonic day 6.5; E10.5, embryonic day10.5; E12.5, embryonic day 12.5.

The erasure of DNA methylation in PGCs is probably the result of a defect in maintenance of DNA methylation, caused by the diminished expression of UHRF1 in the cells [4]. After the demethylation, DNMT3A establishes the methylation pattern in combination with DNMT3L, which itself does not possess enzymatic activity but is indispensable for the activity of DNMT3A [5–7] in oocytes arrested at an early stage of the first meiotic division or in prospermatogonia arrested at the G1 phase [8]. Although the major role of UHRF1 is in the maintenance of DNA methylation (Section 2.2), our group has recently found that UHRF1 is involved in 25% of the genome-wide *de novo* DNA methylation in oocytes [9]. The absence of the UHRF1 protein preferentially decreased DNA methylation levels at transcriptionally inactive regions without histone H3 trimethylation at lysine 36 (H3K36me3) mark. Given that only a small percentage decrease in DNA methylation was observed in *DNMT1* KO oocytes [10] and that UHRF1 has the potential to interact with *de novo* DNMTs [11], UHRF1 may cooperate with DNMT3A for the establishment of methylation patterns. Despite the involvement of UHRF1 in *de novo* DNA methylation in oocytes, our group found that the localization of UHRF1 in oocytes is mainly in the cytoplasm [9]. Recently, cytoplasmic Stella (also known as DPPA3 and PGC7), which is localized in both the cytoplasm and the nucleus, is reported to contribute to the cytoplasmic localization of UHRF1 in oocytes to prevent aberrantly excessive *de novo* DNA methylation by the UHRF1 protein complex [12]. Nuclear Stella is also reported to inhibit the association of UHRF1 with chromatin, resulting in a possible double-layer mechanism to prevent aberrant *de novo* DNA methylation by the complex [13].

During post-implantation embryogenesis and early childhood, not only DNMT3A but also DNMT3B proves to be essential for establishing the characteristic methylation pattern [14]. These enzymes may work together or independently to establish specific DNA methylation patterns in each cell type. However, it still has to be determined when the establishment of the methylation pattern is completed, although it probably depends on the cell type. The "developmental origins of health and disease" (DOHaD) is a concept that has emerged over the past three decades, linking the risk of diseases in later childhood and adult life with the environmental conditions of the early life, including nutrient availability to the mothers. Accumulating evidence suggests that the environment can change the epigenetic state, including DNA methylation of the fetus and infant, with the state being maintained throughout the lifetime of the individual [15]. A well-known experiment showed that early experience in childhood permanently alters behavior and physiology; interactions between rat mothers and their offspring, including the licking and grooming of the pups by their mother in the first week of life, altered the DNA methylation status of the *glucocorticoid receptor* promoter in the hippocampus of the offspring, resulting in differential stress tolerance among the offspring [16]. This indicates that the establishment of DNA methylation is not complete by the first week after birth, at least in the hippocampal neurons of the rat.

2.2 Specification of *de novo* DNA methylation sites

The mechanisms underlying the specification of the genomic regions targeted by *de novo* DNMTs have remained largely elusive. In oocytes, a significant positive correlation between transcription and highly methylated regions has been reported [17]. It is known that transcriptionally active regions are marked with H3K36me3 and that the histone methyltransferase SET domain containing 2 (SETD2) is responsible for the histone methylation in oocytes [18]. Since SETD2 is reported to interact with the phosphorylated C-terminal domain of RNA polymerase II (RNA pol II) [19], SETD2 appears to methylate histones at regions actively transcribed

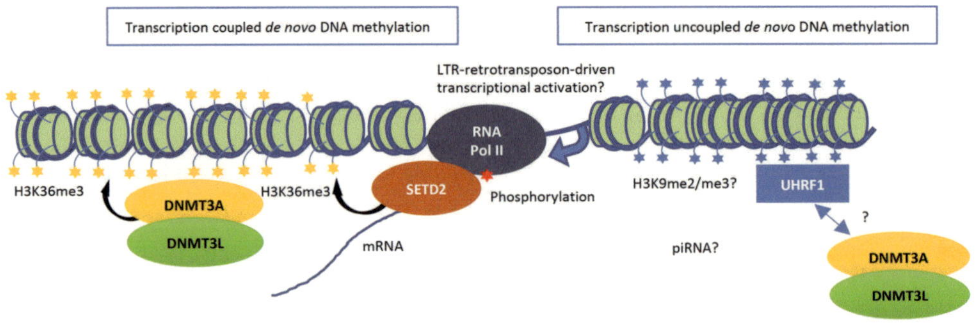

Figure 2.
A model for transcription-coupled and transcription-uncoupled de novo DNA methylation in oocytes. SETD2 methylates H3K36 accompanied with transcription by RNA polymerase II, and DNMT3A recognizes the histone mark and methylates DNA, resulting in the establishment of DNA methylation pattern specific to oocytes. Long terminal repeat (LTR)-retrotransposons activated in PGCs may be partially involved in triggering the transcription. Although UHRF1 is involved in the de novo DNA methylation of regions without the H3K36me3 mark, there could be additional mechanisms for transcription-uncoupled de novo DNA methylation.

by the polymerase. On the other hand, the PWWP domain of DNMT3A recognizes H3K36me3 [20], and mutations in this domain, which disrupt this recognition, cause microcephalic dwarfism with aberrant DNA methylation in humans and in a mouse model [21, 22]. Oocyte-specific *SETD2* KO also causes aberrant DNA methylation [23]. Taken together, it appears that SETD2 methylates H3K36 accompanied by transcription by RNA pol II and DNMT3A recognizes the histone mark and methylates the DNA, resulting in the establishment of DNA methylation patterns specific to oocytes (**Figure 2**). However, there are exceptions. For example, as described above, UHRF1 is involved in 25% of the genome-wide *de novo* DNA methylation, mostly at transcriptionally inactive regions lacking the H3K36me3 mark [9]. It is still unknown which factors trigger transcription in oocytes, although transcription from long terminal repeat (LTR)-retrotransposons, whose methylation is erased in PGCs, could be one such trigger [24].

During embryogenesis, transcription factors probably define certain transcribed regions in each cell type as only four transcriptional factors (OCT3/4, SOX2, KLF4, and MYC), together known as OSKM or Yamanaka factors, can drive drastic transcriptional change and define epigenetically active regions in differentiated cells, resulting in induced pluripotent stem (iPS) cells [25]. DNMTs can access regions, where the transcription factors are absent, to passively specify regions for DNA methylation (**Figure 3**). Noncoding RNAs, such as PIWI-interacting RNAs (piRNAs) and long noncoding RNAs (lncRNAs), can also contribute to the specification of regions for DNA methylation (**Figure 3**). piRNAs are the largest class (26–31 nucleotides) of small noncoding RNA expressed in animal cells, which were first discovered in *Drosophila* as RNAs interacting with the PIWI protein; human and mouse homologs are HIWI and MIWI, respectively. In most cases, precursor piRNAs are derived from piRNA clusters in the genome composed of mutated TEs. The precursor piRNAs are processed by several steps and matured by the addition of a methyl group at their 3′ ends [26]. Then, the maturated piRNAs interact with Argonaute (AGO) family proteins and cleave the TEs, which are undesirably transcribed by the erasure of DNA methylation in PGCs [26]. Although the underlying mechanisms are unknown, piRNAs silence these TEs by epigenetic modifications, including DNA methylation, especially during spermatogenesis [27]. In addition, lncRNAs can specify *de novo* DNA methylation-acquired regions. *X-inactive specific transcript* (*XIST*) is one of the best-studied lncRNAs. *XIST* RNA is randomly expressed from one of two X-chromosomes in mammalian female cells during

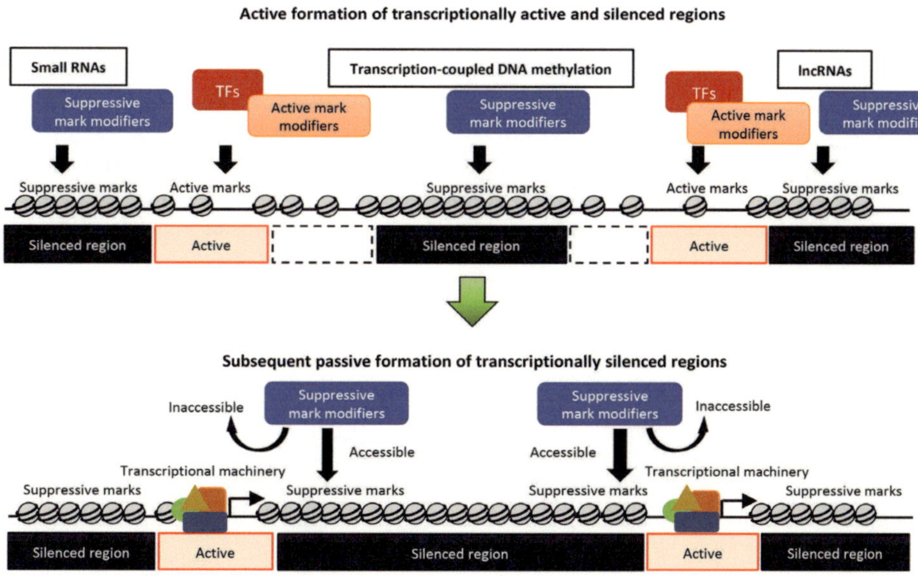

Figure 3.
A model for formation of transcriptionally active and silenced regions during embryogenesis. Transcription factors (TFs) could define transcriptionally active regions, while small RNAs and lncRNAs could define transcriptionally silenced regions in addition to transcription-coupled de novo DNA methylation. After transcriptionally active and silenced regions could be actively determined, suppressive mark modifiers, such as de novo DNMTs, may add suppressive epigenetic marks to accessible regions, which transcriptional machineries do not occupy, resulting in the passive formation of transcriptionally silenced regions.

embryogenesis and covers the X-chromosome *in cis* to trigger silencing of most genes on it by several layers of epigenetic modifications, including DNA methylation, to achieve dosage compensation [28, 29].

3. Maintenance of DNA methylation

3.1 Maintenance of DNA methylation by the DNMT1/UHRF1 complex

Once DNA methylation patterns specific to each cell type are established, the pattern is maintained by the DNMT1/UHRF1 complex throughout the individual's lifetime [2]. UHRF1 (also known as Np95 or ICBP90) is a multidomain protein, which contains a ubiquitin-like (UBL) domain, a tandem Tudor domain (TTD), a plant homeodomain (PHD) finger, a SET and RING-associated (SRA) domain, and a really interesting new gene (RING) domain. The TTD recognizes di–/tri-methylated H3K9 (H3K9me2/me3) and also LIG1 (LIG1K126me2/me3) [30, 31], the PHD recognizes the unmethylated N-terminus of histone H3 and LIG1 [32], the SRA domain recognizes hemi-methylated DNA at the replication fork [33–35], and the RING domain mono-ubiquitylates multiple lysines of histone H3 at K14, K18, and K23 and those of the PCNA-associated factor 15 (PAF15) at K15 and K24 [36–39]. The UBL domain facilitates both the RING-mediated ubiquitylation and the SRA-mediated recognition of hemi-methylated DNA [40, 41].

Current consensus has it that the process of maintenance of DNA methylation operates as follows. After DNA replication, UHRF1 directly recognizes hemi-methylated DNA and mono-ubiquitylates histone H3K14, K18, and K23, to recruit DNMT1 to the hemi-methylation sites. Then, DNMT1 recognizes two of the three ubiquitylated histone lysine residues through the replication foci targeting sequence (RFTS)

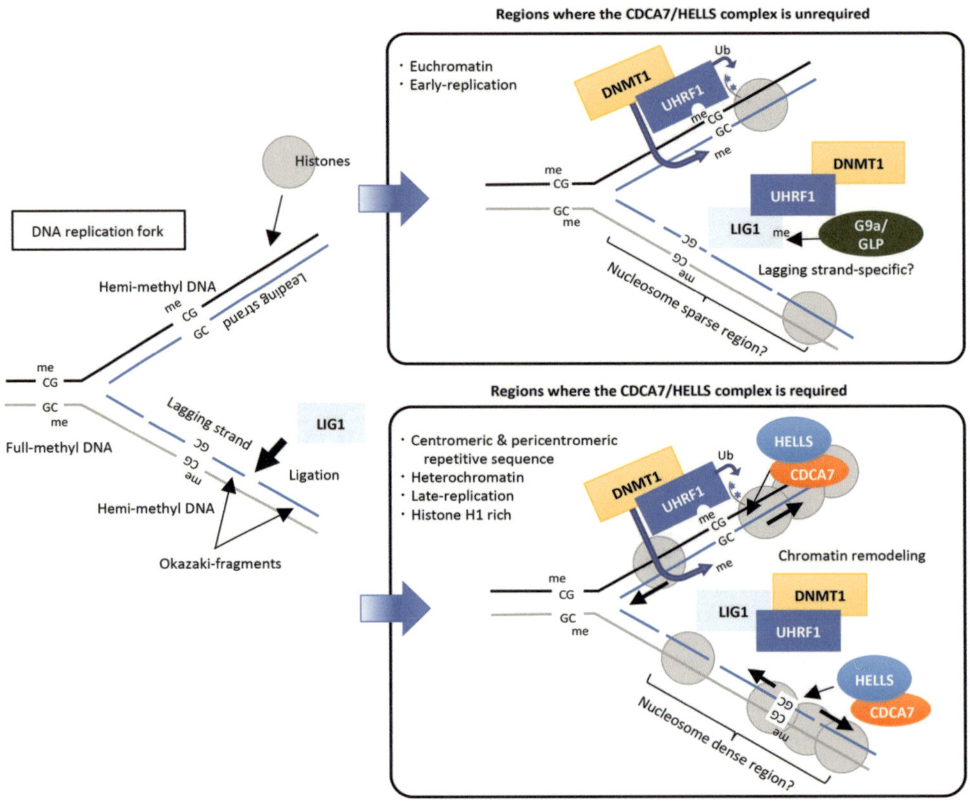

Figure 4.
A model of maintenance of DNA methylation where the CDCA7/HELLS complex is unrequired or required. After DNA replication, UHRF1 directly recognizes hemi-methylated DNA, and mono-ubiquitylates multiple histone lysines, H3K14, K18, and K23, to recruit DNMT1 to the hemi-methylation sites. Consequently, DNMT1 recognizes two of the three ubiquitylated lysine residues through the replication foci targeting sequence (RFTS) domain and methylates the nascent strand in hemi-methylated DNA, resulting in the maintenance of the methylation pattern. LIG1 methylated by G9a/GLP helps UHRF1 to maintain DNA methylation. In addition, the CDCA7/HELLS complex is required for maintaining the DNA methylation of centromeric and pericentromeric regions. The complex may also be required for maintaining regions that are heterochromatic, late replicating, and histone H1 rich, all these regions being nucleosome dense.

domain and methylates the nascent strand in hemi-methylated DNA, resulting in the maintenance of the methylation patterns (**Figure 4**). Immediately prior to the methylation of hemi-methylated DNA by DNMT1, it has been reported that the deubiquitylation of histones by ubiquitin specific peptidase 7 (USP7) is required [42]. DNA ligase 1 (LIG1), which is critical for the joining together of Okazaki fragments [43], is also involved in this process [31]. Euchromatic histone lysine methyltransferase 2 (EHMT2, also called G9a) and EHMT1 (also called GLP) methylate K126 of LIG1. UHRF1 recognizes the methylated LIG1, and this interaction facilitates the recruitment of UHRF1 to DNA replication sites. Since LIG1 is indispensable for completing the lagging strand synthesis, the interaction between UHRF1 and LIG1 may be especially important for maintenance of DNA methylation of the strand (**Figure 4**).

3.2 Maintenance of DNA methylation by the CDCA7/HELLS chromatin remodeling complex

The cell division cycle-associated 7 (CDCA7)/helicase lymphoid-specific (HELLS) chromatin remodeling complex is also involved in maintenance of DNA methylation. Recently, an international group including us identified *CDCA7* and *HELLS* (also known as *LSH*) to be causative genes of the immunodeficiency, centromeric instability, facial anomalies (ICF) syndrome type-3 and type-4 (hereafter ICF3 and ICF4), respectively [44]. The syndrome is a rare autosomal recessive disorder characterized by reduced immunoglobulin levels in the serum and recurrent infection [45]. Centromeric instability manifests as stretched heterochromatin, chromosome breaks, and multiradial configurations involving the centromeric/pericentromeric regions of chromosomes 1, 9, and 16 in activated lymphocytes [46], and the cytological defects are accompanied by DNA hypomethylation in pericentromeric satellite-2 and -3 repeats of these chromosomes.

Patients with the ICF syndrome are classified into two groups [47]. One group includes ICF syndrome type-1 (ICF1), which shows DNA hypomethylation only at the pericentromeric repeats. A causative gene for this group is *DNMT3B* [1, 48, 49]. The second group includes ICF syndrome type-2, type-3, and type-4 (ICF2, ICF3, and ICF4, respectively), which shows DNA hypomethylation at centromeric α-satellite repeats in addition to the pericentromeric repeats. As described above, causative genes for ICF3 and ICF4 are *CDCA7* and *HELLS,* respectively [44]. The causative gene for ICF2 is *zinc finger and BTB domain containing 24* (*ZBTB24*) [50]. As ZBTB24 is a transcriptional activator of *CDCA7* [51, 52], and CDCA7 and HELLS constitute a chromatin remodeling complex, in which CDCA7 stimulates the nucleosome remodeling activity of HELLS [53], the same pathway seems to be disrupted in ICF2, ICF3, and ICF4.

A recent study revealed that, in addition to centromeric and pericentromeric repeats, DNA methylation levels of other heterochromatic late-replicating regions are affected in ICF2, ICF3, and ICF4 patients, though not in ICF1 patients [54]. As *UHRF1* KO and *DNMT1* KO cause hypomethylation of the entire genome, including centromeric and pericentromeric repeats [2], the DNMT1/UHRF1 complex is surely essential for maintaining these regions. However, the CDCA7/HELLS complex seems to be required for assisting the DNMT1/UHRF1 complex to methylate hemimethylated DNA, possibly by sliding nucleosomes in a region-specific manner [53]. Supporting this idea, our group detected an interaction between CDCA7 and UHRF1 [55]. Late-replicating regions tend to be heterochromatic regions, where the nucleosome density is high. Therefore, the CDCA7/HELLS chromatin remodeling complex may be required for such regions (**Figure 4**).

Using human embryonic kidney 293 cells, our group reported that *DNMT3B* KO caused a slight decrease in DNA methylation of pericentromeric repeats after 4 months of KO by the CRISPR/Cas9 system, while *CDCA7* KO and *HELLS* KO caused drastic decreases in DNA methylation even after only 2 months [55], indicating that the CDCA7/HELLS chromatin remodeling complex is essential for maintaining the DNA methylation of the repeats, whereas the requirement of DNMT3B for the maintenance is limited in differentiated cells. In the *CDCA7* KO and *HELLS* KO cells, DNA methylation levels of centromeric repeats were also decreased, but the level of decrease was much less than that of pericentromeric repeats. This indicates that the CDCA7/HELLS complex is less essential for maintenance of DNA methylation of centromeric repeats. Because the chromatin structure, density of nucleosomes, and histone variants are different between centromeric and pericentromeric regions, these differences may determine the levels of requirement for the chromatin remodeling complex. In addition, it has been reported that nucleosomes and the linker histone H1 are barriers to access of DNMTs to DNA and that HELLS and deficient in DNA methylation 1 (DDM1), a plant homolog of HELLS, are required for the methylation of DNA wrapped around nucleosomes [56, 57]. Consistent with these reports, the most abundant proteins co-immunoprecipitated with human CDCA7 were histone H1 and core histones in our group's report [55]. The interaction between the CDCA7/HELLS complex and histone H1 may also be a cue to identify regions where the complex is required for maintenance of DNA methylation (**Figure 4**).

3.3 Maintenance of DNA methylation by the proteins associated with multi-locus imprint disorder

It is reported that mutations in genes encoding zinc finger protein 57 (ZFP57) and components of subcortical maternal complex (SCMC), including NLRP2, NLRP5, NLRP7, PADI6, OOEP, and TLE6, cause the multi-locus imprint disorder, which exhibits DNA hypomethylation at multiple imprinting control regions (ICRs) [58–61]. Since the hypomethylation is observed in both paternally and maternally methylated ICRs, these factors are thought to be involved in maintenance of DNA methylation against genome-wide DNA demethylation in preimplantation embryos (**Figure 1**). Mutations in *ZFP57* cause transient neonatal diabetes mellitus [61]. As ZFP57 is a nuclear protein, which recognizes the methylated TGCCGC hexanucleotide found in almost all ICRs and which acts together with ZNF445, KRAB-associated protein-1 (KAP1), DNMTs, SET domain bifurcated histone lysine methyltransferase 1 (SETDB1), and heterochromatin protein 1 (HP1) [62, 63], ZFP57 is considered to maintain DNA methylation by directly binding to ICRs with such proteins. However, the mechanism by which SCMC components, which are localized adjacent to the oocyte membrane, can maintain DNA methylation at ICRs remains elusive [59]. Among the multi-locus imprint disorder cases, just one case, who has a heterozygous mutation (V159 M in isoform 1, V172 M in isoform 2) in the TTD of UHRF1, has been reported [60].

4. Conclusions

I identified UHRF1 as a novel methyl-CpG binding protein in 2004 by biotin-avidin pulldown assay using biotin-labeled methylated DNA mixed with nuclear extracts and subsequent mass spectrometric analysis [64, 65]. Since then, an understanding of the mechanism by which maintenance of DNA methylation is achieved has quickly expanded and deepened, progress that I would never have imagined at that time. When the involvement of UHRF1 in maintenance of DNA methylation was reported [2], the recognition of hemi-methylated DNA by UHRF1 was reported [32, 34, 35], and the ubiquitylation of histone H3 by UHRF1 was reported [36], each time I felt that the mechanism of maintenance of DNA methylation had been resolved. However, the mechanism is more complicated than expected, and more factors could still be involved to assist the DNMT1/UHRF1 complex, depending on context such as replication timing, replication strand, and higher-order chromatin structure. We still cannot take our eyes off advances in this field.

Acknowledgements

This work was supported by JSPS KAKENHI Grant number JP18K06961. I would like to thank Enago (www.enago.jp) for the English language review.

Conflict of interest

The author has declared that no conflict of interest exists.

Thanks

I would like to thank Drs. Kyohei Arita (Yokohama City University), Jafar Sharif (RIKEN), and Atsuya Nishiyama (the University of Tokyo) for involving me in the UHRF1 community and discussing many interesting topics. I would like to thank Dr. Hisato Kobayashi (Nara Medical University) for notifying me of a possible role of LTR-retrotransposons in oocytes. I also would like to thank Profs. Hiroyuki Sasaki (Kyushu University) and Hironori Funabiki (Rockefeller University) for deepening my insights into epigenetics and chromatins. Finally, I would like to thank Prof. Yusuke Nakamura (Japanese Foundation for Cancer Research) for encouraging me with challenging experiments, which resulted in the identification of UHRF1, when I was a PhD student.

Author details

Motoko Unoki
Division of Epigenomics and Development, Medical Institute of Bioregulation, Kyushu University, Fukuoka, Japan

*Address all correspondence to: unokim@bioreg.kyushu-u.ac.jp

References

[1] Okano M, Bell DW, Haber DA, Li E. DNA methyltransferases Dnmt3a and Dnmt3b are essential for de novo methylation and mammalian development. Cell. 1999;**99**:247-257. DOI: 10.1016/S0092-8674(00)81656-6

[2] Sharif J et al. The SRA protein Np95 mediates epigenetic inheritance by recruiting Dnmt1 to methylated DNA. Nature. 2007;**450**:908-912. DOI: 10.1038/nature06397

[3] Saitou M, Kagiwada S, Kurimoto K. Epigenetic reprogramming in mouse pre-implantation development and primordial germ cells. Development. 2012;**139**:15-31. DOI: 10.1242/dev.050849

[4] Kagiwada S, Kurimoto K, Hirota T, Yamaji M, Saitou M. Replication-coupled passive DNA demethylation for the erasure of genome imprints in mice. The EMBO Journal. 2013;**32**:340-353. DOI: 10.1038/emboj.2012.331

[5] Suetake I, Shinozaki F, Miyagawa J, Takeshima H, Tajima S. DNMT3L stimulates the DNA methylation activity of Dnmt3a and Dnmt3b through a direct interaction. The Journal of Biological Chemistry. 2004;**279**:27816-27823. DOI: 10.1074/jbc.M400181200

[6] Bourc'his D, Xu GL, Lin CS, Bollman B, Bestor TH. Dnmt3L and the establishment of maternal genomic imprints. Science. 2001;**294**:2536-2539. DOI: 10.1126/science.1065848

[7] Hata K, Okano M, Lei H, Li E. Dnmt3L cooperates with the Dnmt3 family of de novo DNA methyltransferases to establish maternal imprints in mice. Development. 2002;**129**:1983-1993

[8] Sasaki H, Matsui Y. Epigenetic events in mammalian germ-cell development: Reprogramming and beyond. Nature Reviews Genetics. 2008;**9**:129-140. DOI: 10.1038/nrg2295

[9] Maenohara S et al. Role of UHRF1 in de novo DNA methylation in oocytes and maintenance methylation in preimplantation embryos. PLoS Genetics. 2017;**13**:e1007042. DOI: 10.1371/journal.pgen.1007042

[10] Shirane K et al. Mouse oocyte methylomes at base resolution reveal genome-wide accumulation of non-CpG methylation and role of DNA methyltransferases. PLoS Genetics. 2013;e1003439. DOI: 10.1371/journal.pgen.1003439

[11] Meilinger D et al. Np95 interacts with de novo DNA methyltransferases, Dnmt3a and Dnmt3b, and mediates epigenetic silencing of the viral CMV promoter in embryonic stem cells. EMBO Reports. 2009;**10**:1259-1264. DOI: 10.1038/embor.2009.201

[12] Li Y et al. Stella safeguards the oocyte methylome by preventing de novo methylation mediated by DNMT1. Nature. 2018;**564**:136-140. DOI: 10.1038/s41586-018-0751-5

[13] Du W et al. Stella protein facilitates DNA demethylation by disrupting the chromatin association of the RING finger-type E3 ubiquitin ligase UHRF1. The Journal of Biological Chemistry. 2019;**294**:8907-8917. DOI: 10.1074/jbc.RA119.008008

[14] Moarefi AH, Chedin F. ICF syndrome mutations cause a broad spectrum of biochemical defects in DNMT3B-mediated de novo DNA methylation. Journal of Molecular Biology. 2011;**409**:758-772. DOI: 10.1016/j.jmb.2011.04.050

[15] Goyal D, Limesand SW, Goyal R. Epigenetic responses and the developmental origins of health and

disease. The Journal of Endocrinology. 2019;**242**:T105-T119. DOI: 10.1530/JOE-19-0009

[16] Weaver IC et al. Epigenetic programming by maternal behavior. Nature Neuroscience. 2004;7:847-854. DOI: 10.1038/nn1276

[17] Veselovska L et al. Deep sequencing and de novo assembly of the mouse oocyte transcriptome define the contribution of transcription to the DNA methylation landscape. Genome Biology. 2015;**16**:209. DOI: 10.1186/s13059-015-0769-z

[18] Li C et al. Histone methyltransferase SETD2 is required for meiotic maturation in mouse oocyte. Journal of Cellular Physiology. 2018;**234**:661-668. DOI: 10.1002/jcp.26836

[19] Li M et al. Solution structure of the Set2-Rpb1 interacting domain of human Set2 and its interaction with the hyperphosphorylated C-terminal domain of Rpb1. Proceedings of the National Academy of Sciences of the United States of America. 2005;**102**:17636-17641. DOI: 10.1073/pnas.0506350102

[20] Dhayalan A et al. The Dnmt3a PWWP domain reads histone 3 lysine 36 trimethylation and guides DNA methylation. The Journal of Biological Chemistry. 2010;**285**:26114-26120. DOI: 10.1074/jbc.M109.089433

[21] Heyn P et al. Gain-of-function DNMT3A mutations cause microcephalic dwarfism and hypermethylation of Polycomb-regulated regions. Nature Genetics. 2019;**51**:96-105. DOI: 10.1038/s41588-018-0274-x

[22] Sendzikaite G, Hanna CW, Stewart-Morgan KR, Ivanova E, Kelsey G. A DNMT3A PWWP mutation leads to methylation of bivalent chromatin and growth retardation in mice. Nature Communications. 2019;**10**:1884. DOI: 10.1038/s41467-019-09713-w

[23] Xu Q et al. SETD2 regulates the maternal epigenome, genomic imprinting and embryonic development. Nature Genetics. 2019;**51**:844-856. DOI: 10.1038/s41588-019-0398-7

[24] Brind'Amour J et al. LTR retrotransposons transcribed in oocytes drive species-specific and heritable changes in DNA methylation. Nature Communications. 2018;**9**:3331. DOI: 10.1038/s41467-018-05841-x

[25] Takahashi K, Yamanaka S. Induction of pluripotent stem cells from mouse embryonic and adult fibroblast cultures by defined factors. Cell. 2006;**126**: 663-676. DOI: 10.1016/j.cell.2006.07.024

[26] Ozata DM, Gainetdinov I, Zoch A, O'Carroll D, Zamore PD. PIWI-interacting RNAs: Small RNAs with big functions. Nature Reviews. Genetics. 2019;**20**:89-108. DOI: 10.1038/s41576-018-0073-3

[27] Itou D et al. Induction of DNA methylation by artificial piRNA production in male germ cells. Current Biology. 2015;**25**:901-906. DOI: 10.1016/j.cub.2015.01.060

[28] Sado T, Brockdorff N. Advances in understanding chromosome silencing by the long non-coding RNA Xist. Philosophical Transactions of the Royal Society of London. Series B, Biological Sciences. 2013;**368**:20110325. DOI: 10.1098/rstb.2011.0325

[29] Sado T, Sakaguchi T. Species-specific differences in X chromosome inactivation in mammals. Reproduction. 2013;**146**:R131-R139. DOI: 10.1530/REP-13-0173

[30] Karagianni P, Amazit L, Qin J, Wong J. ICBP90, a novel methyl K9 H3 binding protein linking protein

ubiquitination with heterochromatin formation. Molecular and Cellular Biology. 2008;**28**:705-717. DOI: 10.1128/MCB.01598-07

[31] Ferry L et al. Methylation of DNA Ligase 1 by G9a/GLP recruits UHRF1 to replicating DNA and regulates DNA methylation. Molecular Cell. 2017;**67**:550-565. DOI: 10.1016/j.molcel.2017.07.012

[32] Arita K et al. Recognition of modification status on a histone H3 tail by linked histone reader modules of the epigenetic regulator UHRF1. Proceedings of the National Academy of Sciences of the United States of America. 2012;**109**:12950-12955. DOI: 10.1073/pnas.1203701109

[33] Arita K, Ariyoshi M, Tochio H, Nakamura Y, Shirakawa M. Recognition of hemi-methylated DNA by the SRA protein UHRF1 by a base-flipping mechanism. Nature. 2008;**455**:818-821. DOI: 10.1038/nature07249

[34] Hashimoto H et al. The SRA domain of UHRF1 flips 5-methylcytosine out of the DNA helix. Nature. 2008;**455**: 826-829. DOI: 10.1038/nature07280

[35] Avvakumov GV et al. Structural basis for recognition of hemi-methylated DNA by the SRA domain of human UHRF1. Nature. 2008;**455**: 822-825. DOI: 10.1038/nature07273

[36] Nishiyama A et al. Uhrf1-dependent H3K23 ubiquitylation couples maintenance DNA methylation and replication. Nature. 2013;**502**:249-253. DOI: 10.1038/nature12488

[37] Qin W et al. DNA methylation requires a DNMT1 ubiquitin interacting motif (UIM) and histone ubiquitination. Cell Research. 2015;**25**:911-929. DOI: 10.1038/cr.2015.72

[38] Ishiyama S et al. Structure of the Dnmt1 reader module complexed with a unique two-mono-ubiquitin mark on histone H3 reveals the basis for DNA methylation maintenance. Molecular Cell. 2017;**68**:350-360. DOI: 10.1016/j.molcel.2017.09.037

[39] Karg E et al. Ubiquitome analysis reveals PCNA-associated factor 15 (PAF15) as a specific ubiquitination target of UHRF1 in embryonic stem cells. Journal of Molecular Biology. 2017;**429**:3814-3824. DOI: 10.1016/j.jmb.2017.10.014

[40] DaRosa PA et al. A bifunctional role for the UHRF1 UBL domain in the control of hemi-methylated DNA-dependent histone ubiquitylation. Molecular Cell. 2018;**72**:753-765. DOI: 10.1016/j.molcel.2018.09.029

[41] Foster BM et al. Critical role of the UBL domain in stimulating the E3 ubiquitin ligase activity of UHRF1 toward chromatin. Molecular Cell. 2018;**72**:739-752. DOI: 10.1016/j.molcel.2018.09.028

[42] Yamaguchi L et al. Usp7-dependent histone H3 deubiquitylation regulates maintenance of DNA methylation. Scientific Reports. 2017;**7**:55. DOI: 10.1038/s41598-017-00136-5

[43] Levin DS, McKenna AE, Motycka TA, Matsumoto Y, Tomkinson AE. Interaction between PCNA and DNA ligase I is critical for joining of Okazaki fragments and long-patch base-excision repair. Current Biology. 2000;**10**:919-922. DOI: 10.1016/S0960-9822(00)00619-9

[44] Thijssen PE et al. Mutations in CDCA7 and HELLS cause immunodeficiency-centromeric instability-facial anomalies syndrome. Nature Communications. 2015;**6**:7870. DOI: 10.1038/ncomms8870

[45] Ehrlich M. The ICF syndrome, a DNA methyltransferase 3B deficiency and immunodeficiency disease. Clinical Immunology. 2003;**109**:17-28. DOI: 10.1016/S1521-6616(03)00201-8

[46] Maraschio P, Zuffardi O, Dalla Fior T, Tiepolo L. Immunodeficiency, centromeric heterochromatin instability of chromosomes 1, 9, and 16, and facial anomalies: The ICF syndrome. Journal of Medical Genetics. 1988;**25**:173-180. DOI: 10.1136/jmg.25.3.173

[47] Jiang YL et al. DNMT3B mutations and DNA methylation defect define two types of ICF syndrome. Human Mutation. 2005;**25**:56-63. DOI: 10.1002/humu.20113

[48] Xu GL et al. Chromosome instability and immunodeficiency syndrome caused by mutations in a DNA methyltransferase gene. Nature. 1999;**402**:187-191. DOI: 10.1038/46052

[49] Hansen RS et al. The DNMT3B DNA methyltransferase gene is mutated in the ICF immunodeficiency syndrome. Proceedings of the National Academy of Sciences of the United States of America. 1999;**96**:14412-14417. DOI: 10.1073/pnas.96.25.14412

[50] de Greef JC et al. Mutations in ZBTB24 are associated with immunodeficiency, centromeric instability, and facial anomalies syndrome type 2. American Journal of Human Genetics. 2011;**88**:796-804. DOI: 10.1016/j.ajhg.2011.04.018

[51] Wu H et al. Converging disease genes in ICF syndrome: ZBTB24 controls expression of CDCA7 in mammals. Human Molecular Genetics. 2016;**25**:4041-4051. DOI: 10.1093/hmg/ddw243

[52] Thompson JJ et al. ZBTB24 is a transcriptional regulator that coordinates with DNMT3B to control DNA methylation. Nucleic Acids Research. 2018;**46**:10034-10051. DOI: 10.1093/nar/gky682

[53] Jenness C et al. HELLS and CDCA7 comprise a bipartite nucleosome remodeling complex defective in ICF syndrome. Proceedings of the National Academy of Sciences of the United States of America. 2018;**115**:E876-E885. DOI: 10.1073/pnas.1717509115

[54] Velasco G et al. Comparative methylome analysis of ICF patients identifies heterochromatin loci that require ZBTB24, CDCA7 and HELLS for their methylated state. Human Molecular Genetics. 2018;**27**:2409-2424. DOI: 10.1093/hmg/ddy130

[55] Unoki M, Funabiki H, Velasco G, Francastel C, Sasaki H. CDCA7 and HELLS mutations undermine nonhomologous end joining in centromeric instability syndrome. The Journal of Clinical Investigation. 2019;**129**:78-92. DOI: 10.1172/JCI99751

[56] Zemach A et al. The Arabidopsis nucleosome remodeler DDM1 allows DNA methyltransferases to access H1-containing heterochromatin. Cell. 2013;**153**:193-205. DOI: 10.1016/j.cell.2013.02.033

[57] Lyons DB, Zilberman D. DDM1 and Lsh remodelers allow methylation of DNA wrapped in nucleosomes. eLife. 2017;**6**:e30674. DOI: 10.7554/eLife.30674

[58] Li X et al. A maternal-zygotic effect gene, Zfp57, maintains both maternal and paternal imprints. Developmental Cell. 2008;**15**:547-557. DOI: 10.1016/j.devcel.2008.08.014

[59] Monk D, Sanchez-Delgado M, Fisher R. NLRPs, the subcortical maternal complex and genomic imprinting. Reproduction. 2017;**154**:R161-R170. DOI: 10.1530/REP-17-0465

[60] Begemann M et al. Maternal variants in NLRP and other maternal effect proteins are associated with multilocus imprinting disturbance in offspring. Journal of Medical Genetics. 2018;**55**:497-504. DOI: 10.1136/jmedgenet-2017-105190

[61] Mackay DJ et al. Hypomethylation of multiple imprinted loci in individuals with transient neonatal diabetes is associated with mutations in ZFP57. Nature Genetics. 2008;**40**:949-951. DOI: 10.1038/ng.187

[62] Quenneville S et al. In embryonic stem cells, ZFP57/KAP1 recognize a methylated hexanucleotide to affect chromatin and DNA methylation of imprinting control regions. Molecular Cell. 2011;**44**:361-372. DOI: 10.1016/j.molcel.2011.08.032

[63] Takahashi N et al. ZNF445 is a primary regulator of genomic imprinting. Genes & Development. 2019;**33**:49-54. DOI: 10.1101/gad.320069.118

[64] Unoki M, Nishidate T, Nakamura Y. ICBP90, an E2F-1 target, recruits HDAC1 and binds to methyl-CpG through its SRA domain. Oncogene. 2004;**23**:7601-7610. DOI: 10.1038/sj.onc.1208053

[65] Unoki M, Nakamura Y. Methylation at CpG islands in intron 1 of EGR2 confers enhancer-like activity. FEBS Letters. 2003;**554**:67-72. DOI: 10.1016/S0014-5793(03)01092-5

Section 3

Methylation Mechanisms in Process of Cancer

Chapter 5

DNA Hydroxymethylation in the Regulation of Gene Expression in Human Solid Cancer

Sofia L. Alcaraz-Estrada, Gabriela Leija-Montoya, Nicolás Serafín-Higuera, Silvia García, Claudia E. Millán-Testa, Mónica Sierra-Martínez, Magali Blanco-Morales and Jorge Sandoval-Basilio

Abstract

Chromatin in cancer undergoes chemical and structural changes that alter gene expression patterns. One of the chemical modifications that impacts gene regulation is 5-hydroxymethylcytosine (5hmC), also called DNA hydroxymethylation. 5hmC is a stable mark that is commonly associated with transcriptional activation. In cancer, the global loss of 5hmC is a hallmark. In addition, the deregulation of 5hmC in specific regions of the genome, such as enhancers, promoters, and body of the gene, alters the expression of genes in cancer. These alterations have been detected by the improvement in the mapping of 5hmC at genomic scale, which has allowed us to evaluate the sites where 5hmC alterations occur and the genes that are affected. In this chapter, the recent knowledge about the status of 5hmC in genome specific sites of human solid cancers, the relationship with enzymes ten-eleven translocation (TET) and **isocitrate dehydrogenase** (IDH) involved in the dynamic regulation of 5hmC levels, and the impact of the 5hmC aberrant changes on the genic expression in these malignances is reviewed.

Keywords: DNA hydroxymethylation, 5hmC, cytosine modifications, DNA demethylation, epigenetics, chromatin, gene expression, cancer

1. Introduction

In carcinogenesis, genetic alterations are necessary along with the deregulation of the epigenetic phenomena. Epigenetics could be defined as the study of the mechanisms that control gene expression without modifying the DNA sequence [1]. In cancer, epigenetic changes can be used to identify the site of origin of the tumor, detect malignant tumors in the earliest stages, and also allow the identification of more aggressive tumors and predict the response to drug therapy [2, 3]. On the other hand, they can be used as therapeutic targets in epigenetic therapy [4].

DNA methylation is a widely studied epigenetic phenomenon, and it occurs predominantly (80%) in a CpG context, where cytosine (C) is methylated in carbon 5, generating 5-methylcytosine (5mC). 5mC constitutes approximately 1% of all

IntechOpen

DNA bases [5] and is associated with the regulation of gene expression, recruitment of transcription factors, nucleosome positioning, splicing, and imprinting. Specifically, in cancer, the overall decrease of 5mC is a general brand and is enriched in specific areas such as the promoter and the first exon, which generates gene silencing and is associated with the development of cancer [6]. Although 5mC is a stable mark, 5mC was shown to oxidize to 5-hydroxymethylcytosine (5hmC). In mammals, the conversion of 5mC to 5hmC is catalyzed by members of the ten-eleven translocation family (TET1, TET2, and TET3), α-ketoglutarate (αKG), and Fe^{2+}-dependent dioxygenases. TET enzymes also oxidize 5hmC to 5-formylcytosine (5fC) and 5-carboxycytosine (5cC) [7].

Currently, it is reported that DNA hydroxymethylation is a stable epigenetic mark that the cells can inherit to its daughter's cells rather than just a transition state [8]. This mark represents the second most abundant C variant in the mammalian genome, always at levels below 5mC, but on the other hand, the 5hmC is 10–100 times higher than the 5fC and 5cC [9, 10].

In mammals, 5hmC occurs almost always (99.89%) in a CpG context [11] and, interestingly, is more enriched in distal regulatory elements of the promoter (46.4%), the body of the gene, and near the cis elements of transcription factors and is less abundant in the promoter region [9–11], suggesting that the hydroxymethylation of DNA has an extensive function in gene regulation. In addition, the enrichment of 5hmC in the different functional elements has been associated with gene transcriptional activation [9].

In humans, the presence of 5hmC would vary significantly between tissues: in the brain (0.67%), rectum (0.57%), liver (0.46%), colon (0.45%), and kidney (0.38%), the 5hmC levels are high, while in the lung (0.14%), they are relatively low and very low in the heart (0.05%), breast (0.05%), and placenta (0.06%) [12]. Compared to normal tissue, many solid neoplasms (e.g., breast, colon, prostate, and melanoma) are characterized by the overall loss of 5hmC. In some isolated cases of cancer, it was reported that 5hmC usually increases; however, the general trend is a global decrease of 5hmC in carcinogenesis [13]. Importantly, hypo-5hmC in cancer occurs in the body of the gene, enhancers, and near cis elements of transcription factors, altering gene expression [14, 15]. However, so far, the role of hydroxymethylation of DNA in cancer biology is not completely clear, and more studies are needed that provide deeper information on functions or potential applications as biomarkers. The purpose of this chapter is to provide current knowledge of the deregulation of 5hmC in genome specific sites, the relationship with enzymes ten-eleven translocation (TET) and **isocitrate dehydrogenase** (IDH) involved in the dynamic regulation of 5hmC levels, and its impact on gene expression in different human cancers.

2. 5hmC status in solid cancer

2.1 5hmC status in melanoma

Melanoma is a type of melanocyte neoplasm that is considered highly aggressive [16]. As one of the most aggressive human tumors, it can perform distal and lethal metastases despite the volume of the tumor being 1 mm^3 [17].

Melanoma is a complex disease influenced by genetic and epigenetic alterations. Importantly, epigenetic phenomena in this tumor include hypermethylation of the phosphatase and tensin homolog (PTEN) promoter and $p16^{ink4}$, which is associated with the silencing of tumor suppressor genes [18, 19]. In addition, repressive chromatin marks that silence the TGF-pathway have been reported [20]. On the other hand, the increase in chromatin-modifying enzymes, such as Ezh2 methyltransferase, has been published [21]. A significant decrease in members of the TET enzyme family has been reported, which correlates with low global levels of 5hmC [17–22]. Thus, altered patterns of 5hmC in melanomas have been observed, and in this sense, the analysis of 5hmC in melanomas showed a hypo-hydroxymethylation in the body of the Ras-related C3 botulinum toxin substrate 3 (RAC3) gene, the type 1 insulin-like growth factor receptor (IGF1R), and tissue inhibitor of metalloproteinases 2 (TIMP2) (**Table 1**). However, the effect at the expression level was not determined. In two studies published independently, they reported that in melanoma the expression levels of IGF1R and TIMP2 are high [23, 24], suggesting that hypo-5hmC in the body of the gene probably contributes to the high expression of the IGF1R and TIMP2 gene. On the other hand, it has been observed that the overexpression of TIMP2 in B16F10 melanoma cells reduces invasion and angiogenesis and inhibits apoptosis [24].

2.2 5hmC status in glioma cancer

Gliomas are the most common brain tumors, being classified by grades (I–IV), based on differentiation status, malignant potential, response to treatment, and patient survival rate. Grades III and IV are referred as high-grade glioma and have the worst prognosis with a median survival for grade III of 2–3 years, while grade IV, named glioblastoma (GBM), has approximately 15 months [45]. In 2016, the

Name of gene	5hmC status	5hmC variable position	Effect	Target cancer	Refs.
RAC3, IGF1R, TIMP2	Hypo-5hmC	Gene body	ND	Melanoma	[17]
SOX2-OT, CHD2	Hyper-5hmC	TSS1500	Activation	Glioma	[25]
LSMEM1	Hyper-5hmC	5′UTR	Activation	Glioma	[25]
v-myc, FAM49A, DDX1, IL-2, IL-15, PRC2	Hyper-5hmC	ND	Activation	Neuroblastoma	[26, 27]
PTEN, hMLH1, IRX1	Hypo-5hmC	Promoter	Repression	Gastric	[28]
GATA6, MMP11, VAV2, LATS2	Hyper-5hmC	Promoter and gene body	Activation	Pancreas	[29]
TBX15	Hyper-5hmC	Promoter	Activation	HCC	[30]
COMT, FMO3, LCAT	Hypo-5hmC	Promoter	Repression	HCC	[30]
CCNY, CDK16	Hyper-5hmC	Loci and promoter	Activation	HuRCSC	[31]
VHL, SETD2	Hypo-5hmC	Gene body	Repression	ccRCC	[14]
CA2, FMN2, PDCD4, PKIB, SLC26A2, ALOX15*, GHRHR*, TFP12*, TKTL1*	Hypo-5hmC	Loci, promoter	Repression	Colon	[32–42]
TESC, TGFBI, BMP7, NKD2	Hyper-5hmC	Loci	Activation	Colon	[32]
GLO1	Hyper-5hmC	Promoter	Activation	Endometrial	[43]
LZTS1	Hypo-5hmC	Loci	Repression	Breast	[44]

**The effect of 5hmC was not determined in these genes. 5hmC, 5-hydroxymethylcytosine; TSS1500, 1500 bases upstream transcription start site; 50 UTR, 50 untranslated region; HCC, Hepatocellular carcinoma; HuRCSS, Kidney renal stem cells; ccRCC, Clear cell renal cell carcinoma; ND, not determined.*

Table 1.
Genes with aberrant 5hmC in different solid human neoplasms.

World Health Organization (WHO) introduced a new classification where molecular markers were taken into account. In this classification glioma are divided into subtypes based on the isocitrate dehydrogenase 1 (IDH1) gene mutation status [46]. IDH can affect DNA hypermethylation at certain promoter regions, resulting in a glioma CpG island methylator phenotype [47]. In glioma, it has been described as an aberrant 5mC status in CpG island shores and is 5hmC-dependent, and it correlates with disease progression [15]. Moreover, in a previous report, the 5hmC patterns were analyzed in GBM samples. They observed a 3.5 reduction in the total 5hmC content and of what was present, localized primarily in super-enhancers and cis elements of transcription factors associated with proliferation. Also, they observed a significant enrichment of 5hmC sites in active transcribed genes in GBM. They reported a total of 2121 active transcribed genes of which 146 have the highest proportion of 5hmC.

As an example, genes with hyper-5hmC in 1500 bases upstream of the transcription star site (TSS1500) that are transcriptionally active are SOX2 overlapping transcript (SOX-OT) and chromodomain-helicase-DNA-binding protein 2 (CHD2) (**Table 1**). Additionally, the gene that encodes the leucine-rich single-pass membrane protein 1 (LSMEM1, also called C7orf53) is transcriptionally active and also presented elevated levels of 5hmC in the 5′ untranslated region (5′UTR) (**Table 1**) [25].

2.3 5hmC status in pediatric embryonal tumors

The pediatric embryonal tumors are a rare type of childhood cancers that derive from neuroectodermal tissue and share related histopathological features despite distinct anatomical locations and diverse clinical outcomes [48]. These tumors can originate in many parts of the body, the ones that are derived from the sympathetic nervous system are called neuroblastoma, and the ones that are derived in the brain are called medulloblastoma [49].

2.3.1 Medulloblastoma

Medulloblastoma (MB) is the most common malignant brain tumor of childhood, the overall 5-year disease-free survival remains low (36%) for patients with dissemination, and prognosis remains poor for patients with recurrent MB [50]. In addition, majority of survivors exhibit long-term neurocognitive and neuroendocrine complication as a result of therapy [51, 52].

The analysis of the 5hmC levels in MB showed a reduction, in comparison to non-neoplastic cerebellum [53]. This finding agrees with other reports where the loss of 5hmC is a common event in other brain tumors as well as tumors of different origins [54]. With this discovery, the expression of TET1, TET2, TET3, IDH1, and IDH2 was analyzed, but the profile could not explain the reduction of 5hmC [53], although the difference in expression of this genes did correlate with the different MB molecular subgroups, suggesting a possible role for TET and IDH genes in the control of specifically developmental pathways activated in MB subgroups. This inability to associate the overall reduction of 5hmC levels and the expression of TET and IDH genes could be due to the method of selection of the 5hmC screening and the lack of analysis of the genomic distribution of 5hmC.

2.3.2 Neuroblastoma

Neuroblastoma is accountable for more than 7% of malignancies in patients younger than 15 years and is responsible for 15% of all pediatric oncology deaths.

Risk assessment based on several clinical and biological features, including age, stage, avian myelocytomastosis viral oncogene (v-myc) status, ploidy, and histology, classified the patients into three groups, low-risk (LR), intermediate-risk (IR), and high-risk (HR) disease [26]. Comparison of the 5hmC profile in LR vs. HR permitted the identification of 3320 genes with differential 5hmC levels between the groups. In the LR group, genes with 5hmC enrichment and increased expression were v-myc, family with sequence similarity 49 member A (FAM49A), and DEAD-box helicase 1 (DDX1) (**Table 1**). Regarding the HR group, the genes with hyper-5hmC and high expression include genes involved in inflammation (IL-2 and IL-15) and in the polycomb repressive complex 2 (PRC2) [27].

2.4 5hmC status in parathyroid cancer

Parathyroid cancer is a rare, indolent, and slowly progressive tumor, being the rarest cause of primary hyperparathyroidism. Surgery is the option of treatment; thus the early identification in the preoperative period is vital [55]. Clinical characteristics of parathyroid cancer can overlap with benign parathyroid disease [55, 56]. Barazeghi E. et al. showed reduced global levels of 5hmC in samples of parathyroid carcinoma compared with samples of normal tissues as well as benign parathyroid adenomas; thus, it was suggested that 5hmC level could be a marker to differentiate between benign and malign tumors [56]. Analysis of TET protein expression indicated variable expression of TET1 in parathyroid adenomas and carcinomas, and additionally reduced or absent expression of TET2 was observed in parathyroid carcinomas as compared with normal parathyroid tissue in concordance with the reduced levels of 5hmC reported in parathyroid carcinoma [56, 57]. Furthermore, increased levels of methylation in promoter CpG islands from TET2 were reported [57]. Analysis of genes regulated by hydroxymethylation on this cancer has not been reported, yet.

2.5 5hmC status in thyroid cancer

Thyroid cancer is the most common tumor of the endocrine organs, accounting for 90% of endocrine tumors. In general, thyroid cancer is originated of follicular cells being divided in papillary thyroid carcinoma and follicular thyroid carcinoma; a reduced percent of thyroid cancers is generated from parafollicular C cells being classified as medullary thyroid carcinomas [58]. The information about 5hmC status is extremely limited, but it has been observed that levels of 5hmC are reduced in papillary thyroid carcinomas as compared with control tissue [59]. However, information about expression of TET proteins and regulation by hydroxymethylation or target genes in this cancer or its subtypes is absent.

2.6 5hmC status in oral cancer

Cancer of the oral cavity is the most common tumor worldwide, the squamous cell carcinoma being the most common histopathology type [60]. A decreased expression of TET2 and a reduction of 5mhC levels in samples of oral squamous cell carcinomas have been reported compared with healthy oral tissues by immunohistochemistry [61]. Also, 5hmC levels decreased progressively from benign oral mucosal lesions to oral squamous cell carcinoma [62]. Expression analysis of target genes regulated by 5hmC or mechanisms implicated have not been reported yet.

2.7 5hmC status in gastric cancer

Gastric cancer is the third cause of cancer death. Environment factors, infections, and genetic and epigenetic alterations are related with development of this cancer [63]. Decreased expression of TET1 mRNA and protein has been reported in gastric cancer, while the expression of TET2 and TET3 did not show differences in expression compared with control tissue. In addition, diminished expression of TET1 has been associated with decreased levels of 5hmC in the promoter of PTEN, human mutL homolog 1 (hMLH1), and iroquois homeobox 1 (IRX1), correlating with their reduced expression levels (**Table 1**) [28]. Thus, reduced expression of TET1 and decreased levels of 5hmC in gastric cancer could be related with the decreased expression of suppressor tumors genes. An in-depth analysis of genes regulated by hydroxymethylation on this cancer could allow new therapeutic strategies.

2.8 5hmC status in pancreatic cancer

Pancreatic cancer is a disease with high mortality rate, being the fourth cause of cancer-related deaths in the United States and most developed countries. Different types of pancreatic cancer can rise; however, the subtype termed pancreatic ductal adenocarcinoma (PDAC) is the most common, which accounts for about 85–90% of cases [64, 65]. The absence of early detection methods, delay in diagnosis, and unsuccessful treatments contribute to the high mortality of this cancer [65].

Low global levels of 5hmC have been reported in pancreatic cancer cell lines and in samples of human tumors compared with healthy pancreatic cells [29]. The reduced levels of 5hmC in samples of human pancreatic tumor tissues correlated with the decreased expression of TET1 [66]. A redistribution of 5hmCs was observed in pancreatic cancer, with enrichment in genomic specific regions as promoters and gene body, particularly of the transcriptional factor GATA6, matrix metallopeptidase 11 (MMP11), vav guanine nucleotide exchange factor 2 (VAV2), and large tumor suppressor kinase 2. The enrichment of 5hmC in this genes is in accordance with the increased expression in human samples of pancreatic cancer [29]. Additionally, it has been suggested that pancreatic cancer patients with high GATA6 survive longer so that GATA6 was proposed as a prognosis marker [67]. On the other hand, high levels of MMP11 were associated with poor prognosis of pancreatic cancer patients [68].

2.9 5hmC status in hepatocellular cancer

Globally, hepatocellular carcinoma (HCC) is the sixth most prevalent cancer and the third leading cause of cancer-related death, estimated to cause the death of 500,000–600,000 people per year [69]. The factors that contribute to mortality in HCC are the 5-year recurrence rates standing at 70% after tumor resection and 15–30% posttransplant [70].

Hepatic B virus (HBV) and hepatitis C (HCV) are the main cause of HCC. For HCC related to HBV, epigenetic alterations play vital roles in hepatocarcinogenesis through direct and indirect mechanisms initiated by HBV [69].

Low global levels of 5hmC have been reported in HCC [69]. A redistribution of 5hmCs was observed in HCC, with an enrichment in specific genomic region as promoters. In this same report, a decrease in the levels of hydroxymethylation was observed in the promoter of different genes, this catechol-O-methyltransferase (COMT), the flavin-containing monooxygenase 3 (FMO3), and lecithin-cholesterol acyltransferase (LCAT) [30] (**Table 1**). The gene T-Box transcription factor 15

(TBX15) with enrichment in the 5hmC promoter region was associated with transcriptional activation, while those that presented a decrease of 5hmC were low [30].

2.10 5hmC status in renal cell carcinoma (RCC)

RCC is the most common parenchymal neoplasm in adults. Among urogenital tumors, it is the second with the highest incidence, which represents 2 to 3% of cancer in humans and 80 to 90% of kidney neoplasms. The most common subtypes of RCC are clear cell carcinoma (ccRCC) and papillary carcinoma [31].

Recently, it was reported that kidney cancer and ccRCC have global decreased levels of 5hmC with respect to normal tissue [14]. Paradoxically in kidney renal stem cells (HuRCSC), hyper-5hmC patterns were detected in specific regions such as the cyclin Y (CCNY) promoter and loci and cyclin-dependent kinase-16 (CDK16) (**Table 1**), which was associated with a transcriptional activation. Likewise, the increased levels of TET1 in HuRCSC were detected, which probably explains the increase of 5hmC in HuRCSC [31].

In another study in which ccRCC was analyzed, it was shown that low levels of 5hmC in the body of the von Hippel–Lindau gene (VHL) and SETD2 methyltransferase are associated with low expression [14] (**Table 1**). In ccRCC loss of SETD2 is associated with genomic instability, aberrant transcriptional program, RNA processing defects, and impacts on cell proliferation, differentiation, and cell death [71]. Additionally, it has been observed that the IDH1 enzyme was significantly downregulated in ccRCC compared to normal kidney cells. Thus, in ccRCC the reduction of IDH1 can be a mechanism for the loss of 5hmC through the downregulation of 2-keto glutarate [14].

2.11 5hmC status in colon cancer

The colon cancer or colorectal cancer is the third most common cancer in the world with more incidence in developed countries [72, 73].

Decreased levels of global hydroxymethylation has been reported in colon cancer [32]. In accordance, decreased expression of TET1 has been reported in samples of human colorectal cancer tissue [33]. Moreover, loss of nuclear expression of TET2 has been observed in colorectal cancer tissue [34]. Thus, aberrant patterns of hydroxymethylation in colorectal cancer has been observed. In this sense, analysis of 5hmC status in colon cancer tissues showed hypo-hydroxymethylation in the promoter region of genes such as:

- Carbonic anhydrase 2 (CA2), which belongs to a group of zinc-binding enzymes, which catalyzes the reversible hydration of CO2 to bicarbonate, that is important for maintenance of pH [32–35]

- Formin 2 (FMN2), involved in cell polarity and cytoskeleton organization and prevents degradation of p21, promoting cell cycle arrest [32–36]

- Programmed cell death 4 (PDCD4), considered a tumor suppressor in colorectal cancer, which can inhibit proliferation and invasion, preventing AP-1 transcription and inhibiting mTOR/Akt [37]

- cAMP-dependent protein kinase inhibitor beta (PKIB), which promotes activation of Akt, contributing to cell growth and proliferation (however, the function of PKIB in colon cancer remains to be determined) [32–39]

- Solute carrier family 26 member 2 (SLC26A2) (low expression of this protein promotes proliferation in vitro observed in colon cancer cell lines) [32–41]
- Arachidonate 15-lipoxygenase (ALOX15)
- Growth hormone-releasing hormone (GHRHR)
- Inhibitor of the tissular factor pathway 2 (TFPI2)
- Transketolase-like protein 1 (TKTL1) (**Table 1**).

Related to 5hmC status and expression, in some cases, the decreased levels of expression of these genes were corroborated. Additionally, hyper-hydroxymethylation and increased expression of different genes have been observed in samples of human primary colon cancer [32–42]. The genes with hyper-5hmC status and high expression include tescalcin (TESC), a calcium-binding protein involved in the promotion of tumorigenesis in colorectal cancer, which activates Akt-dependent NF-κB pathway, promoting proliferation and also contributing to invasion and metastasis in colon cancer [32–42, 74, 75]; transforming growth factor-beta-induced (TGFBI) that promotes cell proliferation, migration, metastasis, and inflammation [32–42, 74–76]; and bone morphogenetic protein 7 (BMP7) that has an augmented expression, but in this case there has been a controversy because in colorectal cancer it has also been reported to have low levels. Additionally, it has been reported that, in colorectal cancer, BMP7 has anticancer activity [32–42, 74–78] and naked cuticle homolog 2 (NKD2) [32].

2.12 5hmC status in endometrial cancer

Endometrial cancer is the most common tumor in the genital tract in developed countries [43]. A report showed a decreased level of 5hmC in samples of endometrial cancer by an ELISA-like reaction. In addition, decreased expression levels of TET1 and TET2 mRNA and increased levels of TET3 mRNA were observed [79]. However, in another report, increased levels of TET1 protein and increased levels of 5hmC in endometrial cancer and hyperplasia using immunohistochemistry were indicated [43]. This discrepancy could be associated to the different levels of regulation of TET1 expression. Interestingly, the authors determined that TET1 could promote the accumulation of 5hmC in the promoter of the glyoxalase I (GLO1) gene (**Table 1**), resulting in the increased expression of GLO1 in endometrial cancer [43]. GLO1 promotes proliferation and chemotherapeutic resistance and contributes to progestin resistance used in the treatment of endometrial cancer [43, 79, 80]. Moreover, it was reported that metformin treatment reduced the expression of TET1 and 5hmC levels, promoting the reduction of GLO1 expression and increasing the sensitivity to progestin in a model in vitro [43].

2.13 5hmC status in breast cancer

Breast cancer (BC) is the most frequent neoplasia in women worldwide, data from the WHO suggests that it comprises 16% of the total of cancer cases. In addition, every year 138 million new cases are detected, and approximately 458,000 deaths occur due to the pathology [81].

In breast cancer, low levels of 5hmC in the locus of the gene leucine zipper putative tumor suppressor (LZTS1) (**Table 1**) have been reported. In addition, the level of LZTS1 expression was low in breast cancer samples compared to normal breast tissue. These results coincide with the low expression of TET1. The results suggest that low levels of 5hmC in the locus of LZTS1 are probably due to the decrease in TET1 [44]. In human cancer, loss of LZTS1 expression has been associated with tumor progression, metastasis, and poor prognosis [82].

3. Conclusion

In the tumors researched in this document, in most of them, an overall global reduction of 5hmC, with accumulation in certain genetic locations or genes, is reported. Some are also related to the overexpression of genes, taking into account that 5hmC is an epigenetic mark of transcription activation. However, the methods used to determine the 5hmC vary between the studies, and not all the methods can be considered reliable to distinguish between 5mC and 5hmC with confidence. Moreover, an area that has not been explored is the effect that 5hmC has on the expression in miRNAs and lncRNA. These observations can be taken to reassess the role of DNA hydroxymethylation status in tumorigenesis.

Expression analyses of TET family members and correlation with 5hmC status have been performed in a wide variety of cancers. However, the biological effect of this aberrant changes in 5hmC levels has not been deeply determined. For example, targeted genes regulated by hydroxymethylation in many cancers have not been established, characterization of the hydroxymethylation patterns in regulator regions of target genes is not determined in some cases, and their expression levels are not corroborated with 5hmC status. Importantly, the effects resulting in the modification or the return to the original state of the hydroxymethylation patterns in cancer are extensively unknown. In addition, since hydroxymethylation can be regulated in different physiological process in health and disease, modifications in 5hmC status could generate undesirable side effects. Thus, more studies are necessary to have a comprehensive understanding of the biological effects and dynamic changes of the hydroxymethylation in cancer, which could allow new therapeutic strategies in the future.

Acknowledgements

We acknowledge support from Universidad Hipócrates and Consejo Nacional de Ciencia y Tecnología CONACyT.

Conflict of interest

The authors declare no conflict of interest.

Appendices and nomenclature

ALOX15	Arachidonate 15-lipoxygenase
BMP7	Bone morphogenetic protein 7
bHlH	Basic helix–loop–helix
C	Cytosine
CA2	Carbonic anhydrase 2
CCNY	Cyclin Y promoter and loci
ccRCC	Clear cell carcinoma cell renal

CHD2	Chromodomain-helicase-DNA-binding protein 2
CDK16	Cyclin-dependent kinase-16
COMT	Catechol O-methyltransferase
CpG	Islands cytokine phophate guanine
DDX1	DEAD-box helicase 1
DNA	Deoxyribonucleic acid
Ezh2	Histone-lysine methyltransferase
FAM49A	Family with sequence similarity 49 member A
Fe2+	Iron 2+
FMN2	Formin 2
FMO3	Flavin-containing monooxygenase 3
GATA6|	GATA binding protein 6
GBM	Glioblastoma
GHRHR	Growth hormone-releasing hormone
GLO1	Glyoxalase I
G9a	Euchromatic histone-lysine N-methyltransferase 2 (EHMT2)
HIF1α	Hypoxia-inducible factor 1 alpha subunit
HIF2α	Hypoxia-inducible factor 2 alpha subunit
hMLH1	Human mutL homolog 1
HR	High-risk
hypo-5hmC	Hypo-5 hydroxymethylcytosine
HuRCSC	Kidney renal stem cell
H3K36me3	Trimethylation of lysine 36 of histone H3
IDH	Isocitrate dehydrogenase
IDH1	Isocitrate dehydrogenase 1
IDH2	Isocitrate dehydrogenase 2
IGF1R	Insulin-like growth factor 1 receptor
lncRNA	Long noncoding RNAs
IR	Intermediate-risk
IRX1	Iroquois homeobox 1
LCAT	Lecithin-cholesterol acyltransferase
LATS2	Suppressor kinase of large tumors 2
LR	Low-risk
LSMEM1	Leucine-rich single-pass membrane protein 1
MB	Medulloblastoma
mm3	Cubic millimeter
mRNA	Messenger RNA
MMP11	Matrix metalloproteinase-1
MRPL50	Mitochondrial ribosomal protein L50
MYCN	Proto-oncogene, bHLH transcription factor
NDRG3	Neuregulin 3
NKD2	Naked cuticle homolog 2
NF-κB	Nuclear factor kappa B
PCTAIRE1	Serine/threonine protein kinase (PCTK1)
PDAC	Pancreatic ductal adenocarcinoma
PDCD4	Programmed cell death 4
PKIB	CAMP-dependent protein kinase inhibitor beta
PPP1R3A	Protein phosphatase 1 regulatory subunit 3A
PRC2	Polycomb repressive complex 2
PTEN	Phosphatase and tensin homolog
p16ink	Protein 16 ink
RAC3	Ras-related C3 botulinum toxin substrate 3
RCC	Renal cell carcinoma

RNA	Ribonucleic acid
SETD2	SET domain containing 2, histone lysine methyltransferase
SLC26A2	The solute carrier family 26 member 2 (diastrophic dysplasia sulfate transporter)
SOX2-OT	SOX2 overlay transcript
TESC	Tescalcin
TET	Ten-eleven translocation
TET1	Ten-eleven translocation 1
TET2	Ten-eleven translocation 2
TET3	Ten-eleven translocation 3
TFPI2	Tissue factor pathway inhibitor 2
TGF	Transforming growth factor
TGFBI	Transforming growth factor-beta-induced
TKTL1	Transketolase-like protein1
TIMP2	Tissue inhibitor of metalloproteinase 2
TRDN	Triadin
v-myc	Avian myelocytomastosis viral oncogene
VAV2	Vav guanine nucleotide exchange factor 2
VHL	Von Hippel–Lindau gene
5fC	5-Formylcytosine
5cC	5-Carboxycytosine
5mC	5-Methylcytosine
5hmC	5-Hydroxymethylcytosine
5′UTR	5′ untranslated region
αKG	α-Ketoglutarate

Author details

Sofia L. Alcaraz-Estrada[1†], Gabriela Leija-Montoya[2†], Nicolás Serafín-Higuera[3], Silvia García[4], Claudia E. Millán-Testa[5], Mónica Sierra-Martínez[6], Magali Blanco-Morales[5] and Jorge Sandoval-Basilio[5*]

1 Genomic Medicine Division, National Medical Center, 20 de Noviembre ISSSTE, Mexico City, Mexico

2 Medical School Campus Mexicali, Autonomous University of Baja California, Mexicali, Mexico

3 Cell Biology Laboratory, Health Sciences Unit, Dentistry Faculty, Autonomous University of Baja California, Mexicali, Mexico

4 Clinical Research Service, National Medical Center, 20 de Noviembre ISSSTE, Mexico City, Mexico

5 Molecular Biology Laboratory, Hipocrates University, Acapulco de Juarez, Guerrero, Mexico

6 Laboratory of Genetics and Molecular Diagnostics, Juarez Hospital of México, Mexico City, Mexico

*Address all correspondence to: jorgesandoval@uhipocrates.edu.mx; sanbajl83@gmail.com

† Both authors have contributed equally to this work.

References

[1] Leija Montoya G, González Ramírez J, Sandoval Basilio J, Serafín Higuera I, Isiordia Espinoza M, González González R, et al. Long non-coding RNAs: Regulators of the activity of myeloid-derived suppressor cells. Frontiers in Immunology [Internet]. 2019;**10**:1734. Available from: https://www.frontiersin.org/article/10.3389/fimmu.2019.01734/full

[2] Moran S, Martínez-Cardús A, Sayols S, Musulén E, Balañá C, Estival-Gonzalez A, et al. Epigenetic profiling to classify cancer of unknown primary: A multicentre, retrospective analysis. The Lancet Oncology [Internet]. October 2016;**17**(10):1386-1395. Available from: http://www.ncbi.nlm.nih.gov/pubmed/27575023 [Cited 24 January 2020]

[3] Kamińska K, Nalejska E, Kubiak M, Wojtysiak J, Żołna Ł, Kowalewski J, et al. Prognostic and predictive epigenetic biomarkers in oncology. Molecular Diagnosis & Therapy [Internet]. 2019;**23**(1):83-95. Available from: http://www.ncbi.nlm.nih.gov/pubmed/30523565 [Cited 24 January 2020]

[4] Morel D, Jeffery D, Aspeslagh S, Almouzni G, Postel-Vinay S. Combining epigenetic drugs with other therapies for solid tumours—Past lessons and future promise. Nature Reviews Clinical Oncology [Internet]. 2020;**17**(2):91-107. Available from: http://www.nature.com/articles/s41571-019-0267-4 [Cited 24 January 2020]

[5] Tahiliani M, Koh KP, Shen Y, Pastor WA, Bandukwala H, Brudno Y, et al. Conversion of 5-methylcytosine to 5-hydroxymethylcytosine in mammalian DNA by MLL partner TET1. Science [Internet]. 2009; **324**(5929):930-935. Available from: http://www.sciencemag.org/cgi/doi/10.1126/science.1170116

[6] Sandoval-Basilio J, González-González R, Bologna-Molina R, Isiordia-Espinoza M, Leija-Montoya G, Alcaraz-Estrada SL, et al. Epigenetic mechanisms in odontogenic tumors: A literature review. Archives of Oral Biology [Internet]. 2018;**87**:211-217. Available from: https://linkinghub.elsevier.com/retrieve/pii/S0003996917304120

[7] Wu X, Zhang Y. TET-mediated active DNA demethylation: Mechanism, function and beyond. Nature Reviews Genetics [Internet]. 2017;**18**(9):517-534. Available from: http://www.nature.com/articles/nrg.2017.33

[8] Bachman M, Uribe-Lewis S, Yang X, Williams M, Murrell A, Balasubramanian S. 5-Hydroxymethylcytosine is a predominantly stable DNA modification. Nature Chemistry [Internet]. 2014;**6**(12): 1049-1055. Available from: http://www.nature.com/articles/nchem.2064

[9] Meng H, Cao Y, Qin J, Song X, Zhang Q, Shi Y, et al. DNA methylation, its mediators and genome integrity. International Journal of Biological Sciences [Internet]. 2015;**11**(5):604-617 Available from: http://www.ijbs.com/v11p0604.htm

[10] Ponnaluri VKC, Ehrlich KC, Zhang G, Lacey M, Johnston D, Pradhan S, et al. Association of 5-hydroxymethylation and 5-methylation of DNA cytosine with tissue-specific gene expression. Epigenetics [Internet]. 2017;**12**(2):123-138. Available from: https://www.tandfonline.com/doi/full/10.1080/15592294.2016.1265713

[11] Yu M, Hon GC, Szulwach KE, Song CX, Zhang L, Kim A, et al. Base-resolution analysis of 5-hydroxymethylcytosine in the mammalian genome. Cell. 2012;**149**(6): 1368-1380

[12] Li W, Liu M. Distribution of 5-Hydroxymethylcytosine in different

human tissues. Journal of Nucleic Acids [Internet]. 2011;**2011**:1-5. Available from: http://www.hindawi.com/journals/jna/2011/870726/

[13] Jeschke J, Collignon E, Fuks F. Portraits of TET-mediated DNA hydroxymethylation in cancer. Current Opinion in Genetics & Development [Internet]. 2016;**36**:16-26. Available from: https://linkinghub.elsevier.com/retrieve/pii/S0959437X16000058

[14] Chen K, Zhang J, Guo Z, Ma Q, Xu Z, Zhou Y, et al. Loss of 5-hydroxymethylcytosine is linked to gene body hypermethylation in kidney cancer. Cell Research [Internet]. 2016; **26**(1):103-118. Available from: http://www.nature.com/articles/cr2015150

[15] Fernandez AF, Bayón GF, Sierra MI, Urdinguio RG, Toraño EG, García MG, et al. Loss of 5hmC identifies a new type of aberrant DNA hypermethylation in glioma. Human Molecular Genetics [Internet]. 2018;**27**(17):3046-3059. Available from: https://academic.oup.com/hmg/article/27/17/3046/5033378

[16] Sarkar D, Leung EY, Baguley BC, Finlay GJ, Askarian-Amiri ME. Epigenetic regulation in human melanoma: Past and future. Epigenetics [Internet]. 2015;**10**(2):103-121. Available from: http://www.tandfonline.com/doi/full/10.1080/15592294.2014.1003746

[17] Lian CG, Xu Y, Ceol C, Wu F, Larson A, Dresser K, et al. Loss of 5-Hydroxymethylcytosine is an epigenetic hallmark of melanoma. Cell [Internet]. 2012;**150**(6):1135-1146. Available from: https://linkinghub.elsevier.com/retrieve/pii/S0092867412010124

[18] Jonsson A, Tuominen R, Grafström E, Hansson J, Egyhazi S. High frequency of p16INK4A promoter methylation in NRAS-mutated cutaneous melanoma. Journal of Investigative Dermatology [Internet]. 2010;**130**(12):2809-2817. Available from: https://linkinghub.elsevier.com/retrieve/pii/S0022202X15346509

[19] Mirmohammadsadegh A, Marini A, Nambiar S, Hassan M, Tannapfel A, Ruzicka T, et al. Epigenetic silencing of the PTEN gene in melanoma. Cancer Research [Internet]. 2006;**66**(13): 6546-6552. Available from: http://cancerres.aacrjournals.org/lookup/doi/10.1158/0008-5472.CAN-06-0384

[20] Fiziev P, Akdemir KC, Miller JP, Keung EZ, Samant NS, Sharma S, et al. Systematic Epigenomic analysis reveals chromatin states associated with melanoma progression. Cell Reports [Internet]. 2017;**19**(4):875-889. Available from: https://linkinghub.elsevier.com/retrieve/pii/S2211124717304552

[21] Zingg D, Debbache J, Schaefer SM, Tuncer E, Frommel SC, Cheng P, et al. The epigenetic modifier EZH2 controls melanoma growth and metastasis through silencing of distinct tumour suppressors. Nature Communications [Internet]. 2015;**6**(1):6051. Available from: http://www.nature.com/articles/ncomms7051

[22] Gambichler T, Sand M, Skrygan M. Loss of 5-hydroxymethylcytosine and ten-eleven translocation 2 protein expression in malignant melanoma. Melanoma Research [Internet]. 2013; **23**(3):218-220. Available from: https://insights.ovid.com/crossref?an=00008390-201306000-00008

[23] Yeh AH, Bohula EA, Macaulay VM. Human melanoma cells expressing V600E B-RAF are susceptible to IGF1R targeting by small interfering RNAs. Oncogene. 2006;**25**(50):6574-6581

[24] Zurac S, Neagu M, Constantin C, Cioplea M, Nedelcu R, Bastian A, et al. Variations in the expression of TIMP1, TIMP2 and TIMP3 in cutaneous melanoma with regression and their

possible function as prognostic predictors. Oncology Letters [Internet]. 2016;**11**(5):3354-3360. Available from: https://www.spandidos-publications.com/10.3892/ol.2016.4391

[25] Johnson KC, Houseman EA, King JE, Von Herrmann KM, Fadul CE, Christensen BC. 5-Hydroxymethylcytosine localizes to enhancer elements and is associated with survival in glioblastoma patients. Nature Communications. 2016;**7**(1):1-11

[26] Maris JM, Hogarty MD, Bagatell R, Cohn SL. Neuroblastoma. The Lancet [Internet]. 2007;**369**(9579):2106-2120. Available from: https://linkinghub.elsevier.com/retrieve/pii/S0140673607609830

[27] Applebaum MA, Barr EK, Karpus J, Nie J, Zhang Z, Armstrong AE, et al. 5-Hydroxymethylcytosine profiles are prognostic of outcome in neuroblastoma and reveal transcriptional networks that correlate with tumor phenotype. JCO Precision Oncology [Internet]. 2019;**3**: 1-12. Available from: http://ascopubs.org/doi/10.1200/PO.18.00402

[28] Pei Y, Tao R, Li J, Su L, Yu B, Wu X, et al. TET1 inhibits gastric cancer growth and metastasis by PTEN demethylation and re-expression. Oncotarget [Internet]. 2016;7(21). Available from: http://www.oncotarget.com/fulltext/8900

[29] Bhattacharyya S, Yu Y, Suzuki M, Campbell N, Mazdo J, Vasanthakumar A, et al. Genome-wide hydroxymethylation tested using the HELP-GT assay shows redistribution in cancer. Nucleic Acids Research [Internet]. 2013;**41**(16):e157-e157. Available from: https://academic.oup.com/nar/article-lookup/doi/10.1093/nar/gkt601

[30] Hlady RA, Sathyanarayan A, Thompson JJ, Zhou D, Wu Q, Pham K, et al. Integrating the Epigenome to identify drivers of hepatocellular carcinoma. Hepatology [Internet]. 2019; **69**(2):639-652. Available from: https://onlinelibrary.wiley.com/doi/abs/10.1002/hep.30211

[31] Si Y, Liu J, Shen H, Zhang C, Wu Y, Huang Y, et al. Fisetin decreases TET1 activity and CCNY/CDK16 promoter 5hmC levels to inhibit the proliferation and invasion of renal cancer stem cell. Journal of Cellular and Molecular Medicine [Internet]. 2019;**23**(2): 1095-1105. Available from: http://doi.wiley.com/10.1111/jcmm.14010 [Cited 6 February 2020]

[32] Chapman CG, Mariani CJ, Wu F, Meckel K, Butun F, Chuang A, et al. TET-catalyzed 5-hydroxymethylcytosine regulates gene expression in differentiating colonocytes and colon cancer. Scientific Reports [Internet]. 2015;**5**(1):17568. Available from: http://www.nature.com/articles/srep17568 [Cited 23 January 2020]

[33] Tian Y, Pan F, Sun X, Gan M, Lin A, Zhang D, et al. Association of TET1 expression with colorectal cancer progression. Scandinavian Journal of Gastroenterology. 2017;**52**(3):312-320

[34] Huang Y, Wang G, Liang Z, Yang Y, Cui L, Liu C-Y. Loss of nuclear localization of TET2 in colorectal cancer. Clinical Epigenetics [Internet]. 2016; **8**(1):9. Available from: http://www.clinicalepigeneticsjournal.com/content/8/1/9 [Cited 23 January 2020]

[35] Viikilä P, Kivelä AJ, Mustonen H, Koskensalo S, Waheed A, Sly WS, et al. Carbonic anhydrase enzymes II, VII, IX and XII in colorectal carcinomas. World Journal of Gastroenterology [Internet]. 2016;**22**(36):8168. Available from: http://www.wjgnet.com/1007-9327/full/v22/i36/8168.htm

[36] Li D-J, Feng Z-C, Li X-R, Hu G. Involvement of methylation-associated silencing of formin 2 in colorectal carcinogenesis. World Journal of

Gastroenterology [Internet]. 2018; **24**(44):5013-5024. Available from: http://www.wjgnet.com/1007-9327/full/v24/i44/5013.htm

[37] Long J, Yin Y, Guo H, Li S, Sun Y, Zeng C, et al. The mechanisms and clinical significance of PDCD4 in colorectal cancer. Gene [Internet]. 2019; **680**:59-64. Available from: https://linkinghub.elsevier.com/retrieve/pii/S0378111918309934

[38] Chung S, Furihata M, Tamura K, Uemura M, Daigo Y, Nasu Y, et al. Overexpressing PKIB in prostate cancer promotes its aggressiveness by linking between PKA and Akt pathways. Oncogene [Internet]. 2009; **28**(32):2849-2859. Available from: http://www.nature.com/articles/onc2009144

[39] Zhang JB, Song W, Wang YY, Liu MG, Sun MM, Liu H. Study on correlation between PKIB and pAkt expression in breast cancer tissues. European Review for Medical and Pharmacological Sciences. 2017;**21**(6): 1264-1269

[40] Inoue A, Okamoto K, Fujino Y, Nakagawa T, Muguruma N, Sannomiya K, et al. B-RAF mutation and accumulated gene methylation in aberrant crypt foci (ACF), sessile serrated adenoma/polyp (SSA/P) and cancer in SSA/P. British Journal of Cancer [Internet]. 2015;**112**(2):403-412. Available from: http://www.ncbi.nlm.nih.gov/pubmed/25314065 [Cited 23 January 2020]

[41] Dimberg LY, Towers CG, Behbakht K, Hotz TJ, Kim J, Fosmire S, et al. A genome-wide loss-of-function screen identifies SLC26A2 as a novel mediator of TRAIL resistance. Molecular Cancer Research: MCR [Internet]. 2017;**15**(4):382-394. Available from: http://www.ncbi.nlm.nih.gov/pubmed/28108622 [Cited 23 January 2020]

[42] Li M, Gao F, Xia Y, Tang Y, Zhao W, Jin C, et al. Filtrating colorectal cancer associated genes by integrated analyses of global DNA methylation and hydroxymethylation in cancer and normal tissue. Scientific Reports [Internet]. 2016;**6**(1):31826. Available from: http://www.nature.com/articles/srep31826

[43] Jiang Y, Chen X, Wei Y, Feng Y, Zheng W, Zhang Z. Metformin sensitizes endometrial cancer cells to progestin by targeting TET1 to downregulate glyoxalase I expression. Biomedicine & Pharmacotherapy [Internet]. 2019;**113**:108712. Available from: https://linkinghub.elsevier.com/retrieve/pii/S0753332219304056

[44] Matthias W, Liou W, Pulverer W, Singer CF, Rappaport-Fuerhauser C, Kandioler D, et al. Cytosine 5-Hydroxymethylation of the LZTS1 gene is reduced in breast cancer. Translational Oncology [Internet]. 2013;**6**(6):715-727. Available from: https://linkinghub.elsevier.com/retrieve/pii/S1936523313800113

[45] Stupp R, Mason WP, van den Bent MJ, Weller M, Fisher B, Taphoorn MJB, et al. Radiotherapy plus concomitant and adjuvant temozolomide for glioblastoma. New England Journal of Medicine [Internet]. 2005;**352**(10):987-996. Available from: http://www.nejm.org/doi/abs/10.1056/NEJMoa043330

[46] Verhaak RGW, Hoadley KA, Purdom E, Wang V, Qi Y, Wilkerson MD, et al. Integrated genomic analysis identifies clinically relevant subtypes of Glioblastoma characterized by abnormalities in PDGFRA, IDH1, EGFR, and NF1. Cancer Cell [Internet]. 2010;**17**(1): 98-110. Available from: https://linkinghub.elsevier.com/retrieve/pii/S1535610809004322

[47] Noushmehr H, Weisenberger DJ, Diefes K, Phillips HS, Pujara K, Berman BP, et al. Identification of a CpG island methylator phenotype that defines a distinct subgroup of glioma. Cancer Cell [Internet]. 2010;**17**(5): 510-522. Available from: https:// linkinghub.elsevier.com/retrieve/pii/ S153561081000108X

[48] Chen X, Pappo A, Dyer MA. Pediatric solid tumor genomics and developmental pliancy. Oncogene [Internet]. 2015;**34**(41):5207-5215. Available from: http://www.nature. com/articles/onc2014474

[49] Kohe SE, Bennett CD, Gill SK, Wilson M, McConville C, Peet AC. Metabolic profiling of the three neural derived embryonal pediatric tumors retinoblastoma, neuroblastoma and medulloblastoma, identifies distinct metabolic profiles. Oncotarget [Internet]. 2018;**9**(13). Available from: http://www.oncotarget.com/fulltext/ 24168

[50] Packer RJ, Gajjar A, Vezina G, Rorke-Adams L, Burger PC, Robertson PL, et al. Phase III study of craniospinal radiation therapy followed by adjuvant chemotherapy for newly diagnosed average-risk medulloblastoma. Journal of Clinical Oncology [Internet]. 2006;**24**(25): 4202-4208. Available from: http:// ascopubs.org/doi/10.1200/JCO.2006. 06.4980

[51] Mulhern RK, Palmer SL, Merchant TE, Wallace D, Kocak M, Brouwers P, et al. Neurocognitive consequences of risk-adapted therapy for childhood medulloblastoma. Journal of Clinical Oncology [Internet]. 2005; **23**(24):5511-5519. Available from: http:// ascopubs.org/doi/10.1200/JCO.2005. 00.703

[52] Ribi K, Relly C, Landolt MA, Alber FD, Boltshauser E, Grotzer MA. Outcome of Medulloblastoma in children: Long-term complications and quality of life. Neuropediatrics [Internet]. 2005;**36**(6):357-365. Available from: http://www.thieme-connect.de/ DOI/DOI?10.1055/s-2005-872880

[53] Bezerra Salomão K, Cruzeiro GAV, Bonfim-Silva R, Geron L, Ramalho F, Pinto Saggioro F, et al. Reduced hydroxymethylation characterizes medulloblastoma while TET and IDH genes are differentially expressed within molecular subgroups. Journal of Neuro-Oncology [Internet]. 2018;**139**(1):33-42. Available from: http://link.springer.c om/10.1007/s11060-018-2845-1

[54] Kudo Y, Tateishi K, Yamamoto K, Yamamoto S, Asaoka Y, Ijichi H, et al. Loss of 5-hydroxymethylcytosine is accompanied with malignant cellular transformation. Cancer Science [Internet]. 2012;**103**(4):670-676. Available from: http://doi.wiley.com/ 10.1111/j.1349-7006.2012.02213.x

[55] Machado W. Parathyroid cancer: A review. Cancers [Internet]. 2019;**11**(11): 1676. Available from: https://www.md pi.com/2072-6694/11/11/1676 [Cited 4 February 2020]

[56] Barazeghi E, Gill AJ, Sidhu S, Norlén O, Dina R, Palazzo FF, et al. 5-Hydroxymethylcytosine discriminates between parathyroid adenoma and carcinoma. Clinical Epigenetics [Internet]. 2016;**8**(1):31. Available from: http://www.clinicalepigeneticsjournal. com/content/8/1/31

[57] Barazeghi E, Gill AJ, Sidhu S, Norlén O, Dina R, Palazzo FF, et al. A role for TET2 in parathyroid carcinoma. Endocrine-Related Cancer [Internet]. 2017;**24**(7):329-338. Available from: https://erc.bioscientifica.com/view/ journals/erc/24/7/329.xml

[58] Anania MC, Di Marco T, Mazzoni M, Greco A. Targeting non-oncogene addiction: Focus on thyroid cancer. Cancers [Internet]. 2020;**12**(1):129.

Available from: https://www.mdpi.com/2072-6694/12/1/129 [Cited 5 February 2020]

[59] Tong M, Gao S, Qi W, Shi C, Qiu M, Yang F, et al. 5-Hydroxymethylcytosine as a potential epigenetic biomarker in papillary thyroid carcinoma. Oncology Letters [Internet]. 2019;**18**(3): 2304-2309. Available from: http://www.spandidos-publications.com/10.3892/ol.2019.10531

[60] Montero PH, Patel SG. Cancer of the oral cavity. Surgical Oncology Clinics of North America [Internet]. 2015;**24**(3):491-508. Available from: https://linkinghub.elsevier.com/retrieve/pii/S1055320715000319

[61] Jäwert F, Hasséus B, Kjeller G, Magnusson B, Sand L, Larsson L. Loss of 5-hydroxymethylcytosine and TET2 in oral squamous cell carcinoma. Anticancer Research [Internet]. 2013; **33**(10):4325-4328. Available from: http://www.ncbi.nlm.nih.gov/pubmed/24122999

[62] Cuevas-Nunez MC, CBF G, Woo S-B, Ramsey MR, Chen XL, Xu S, et al. Biological significance of 5-hydroxymethylcytosine in oral epithelial dysplasia and oral squamous cell carcinoma. Oral Surgery, Oral Medicine, Oral Pathology and Oral Radiology [Internet]. 2018;**125**(1):59.e2-73.e2. Available from: https://linkinghub.elsevier.com/retrieve/pii/S2212440317308453

[63] Fattahi S, Kosari-Monfared M, Golpour M, Emami Z, Ghasemiyan M, Nouri M, et al. LncRNAs as potential diagnostic and prognostic biomarkers in gastric cancer: A novel approach to personalized medicine. Journal of Cellular Physiology [Internet]. 2020; **235**(4):3189-3206. Available from: https://onlinelibrary.wiley.com/doi/abs/10.1002/jcp.29260

[64] Jelski W, Mroczko B. Biochemical diagnostics of pancreatic cancer - present and future. Clinica Chimica Acta [Internet]. 2019;**498**:47-51. Available from: https://linkinghub.elsevier.com/retrieve/pii/S0009898119320017

[65] Qian L, Li Q, Baryeh K, Qiu W, Li K, Zhang J, et al. Biosensors for early diagnosis of pancreatic cancer: A review. Translational Research [Internet]. 2019;**213**:67-89. Available from: https://linkinghub.elsevier.com/retrieve/pii/S1931524419301604

[66] Wu J, Li H, Shi M, Zhu Y, Ma Y, Zhong Y, et al. TET1-mediated DNA hydroxymethylation activates inhibitors of the Wnt/β-catenin signaling pathway to suppress EMT in pancreatic tumor cells. Journal of Experimental & Clinical Cancer Research [Internet]. 2019;**38**(1): 348. Available from: https://jeccr.biomedcentral.com/articles/10.1186/s13046-019-1334-5 [Cited 23 January 2020]

[67] Martinelli P, Carrillo-de Santa Pau E, Cox T, Sainz B, Dusetti N, Greenhalf W, et al. GATA6 regulates EMT and tumour dissemination, and is a marker of response to adjuvant chemotherapy in pancreatic cancer. Gut [Internet]. 2017;**66**(9):1665-1676. Available from: http://gut.bmj.com/lookup/doi/10.1136/gutjnl-2015-311256

[68] Lee J, Lee J, Kim JH. Identification of matrix metalloproteinase 11 as a prognostic biomarker in pancreatic cancer. Anticancer Research. 2019; **39**(11):5963-5971

[69] Liu J, Jiang J, Mo J, Liu D, Cao D, Wang H, et al. Global DNA 5-Hydroxymethylcytosine and 5-formylcytosine contents are decreased in the early stage of hepatocellular carcinoma. Hepatology. 2019;**69**(1): 196-208

[70] Chuang K-H, Whitney-Miller CL, Chu C-Y, Zhou Z, Dokus MK, Schmit S, et al. MicroRNA-494 is a master

epigenetic regulator of multiple invasion-suppressor microRNAs by targeting ten eleven translocation 1 in invasive human hepatocellular carcinoma tumors. Hepatology [Internet]. 2015;**62**(2):466-480. Available from: http://doi.wiley.com/10.1002/hep.27816

[71] González-Rodríguez P, Engskog-Vlachos P, Zhang H, Murgoci A-N, Zerdes I, Joseph B. SETD2 mutation in renal clear cell carcinoma suppress autophagy via regulation of ATG12. Cell Death & Disease [Internet]. 2020;**11**(1): 69. Available from: http://www.ncbi.nlm.nih.gov/pubmed/31988284 [Cited 6 February 2020]

[72] Tiwari A, Saraf S, Jain A, Panda PK, Verma A, Jain SK. Basics to advances in nanotherapy of colorectal cancer. Drug Delivery and Translational Research [Internet]. 2019:1-20. Available from: http://link.springer.com/10.1007/s13346-019-00680-9

[73] Jahanafrooz Z, Mosafer J, Akbari M, Hashemzaei M, Mokhtarzadeh A, Baradaran B. Colon cancer therapy by focusing on colon cancer stem cells and their tumor microenvironment. Journal of Cellular Physiology [Internet]. 2019. Available from: http://doi.wiley.com/10.1002/jcp.29337 [Cited 23 January 2020]

[74] Kang J, Kang YH, Oh BM, Uhm TG, Park SY, Kim TW, et al. Tescalcin expression contributes to invasive and metastatic activity in colorectal cancer. Tumour Biology: The Journal of the International Society for Oncodevelopmental Biology and Medicine [Internet]. 2016;**37**(10): 13843-13853. Available from: http://www.ncbi.nlm.nih.gov/pubmed/27485112 [Cited 23 January 2020]

[75] Kang YH, Han SR, Kim J-T, Lee S-J, Il YY, Min J-K, et al. The EF-hand calcium-binding protein tescalcin is a potential oncotarget in colorectal cancer. Oncotarget [Internet]. 2014; **5**(8):2149-2160. Available from: http://www.ncbi.nlm.nih.gov/pubmed/24811141 [Cited 23 January 2020]

[76] Zhu J, Chen X, Liao Z, He C, Hu X. TGFBI protein high expression predicts poor prognosis in colorectal cancer patients. International Journal of Clinical and Experimental Pathology [Internet]. 2015;**8**(1):702-710. Available from: http://www.ncbi.nlm.nih.gov/pubmed/25755764 [Cited 23 January 2020]

[77] Liu R-X, Ren W-Y, Ma Y, Liao Y-P, Wang H, Zhu J-H, et al. BMP7 mediates the anticancer effect of honokiol by upregulating p53 in HCT116 cells. International Journal of Oncology [Internet]. 2017;**51**(3):907-917. Available from: http://www.ncbi.nlm.nih.gov/pubmed/28731124 [Cited 23 January 2020]

[78] Pellatt AJ, Mullany LE, Herrick JS, Sakoda LC, Wolff RK, Samowitz WS, et al. The TGFβ-signaling pathway and colorectal cancer: Associations between dysregulated genes and miRNAs. Journal of Translational Medicine [Internet]. 2018;**16**(1):191. Available from: http://www.ncbi.nlm.nih.gov/pubmed/29986714 [Cited 23 January 2020]

[79] Ciesielski P, Jóźwiak P, Wójcik-Krowiranda K, Forma E, Cwonda Ł, Szczepaniec S, et al. Differential expression of ten-eleven translocation genes in endometrial cancers. Tumor Biology [Internet]. 2017;**39**(3): 101042831769501. Available from: http://journals.sagepub.com/doi/10.1177/1010428317695017

[80] Zhang Z, Dong L, Sui L, Yang Y, Liu X, Yu Y, et al. Metformin reverses progestin resistance in endometrial cancer cells by downregulating GloI expression. International Journal of Gynecologic Cancer [Internet]. 2011;**21**(2):213-221. Available from: http://ijgc.bmj.com/lookup/doi/10.1097/IGC.0b013e318207dac7 [Cited 05 February 2020]

[81] Bray F, Ferlay J, Soerjomataram I, Siegel RL, Torre LA, Jemal A. Global cancer statistics 2018: GLOBOCAN estimates of incidence and mortality worldwide for 36 cancers in 185 countries. CA: A Cancer Journal for Clinicians [Internet]. 2018;**68**(6): 394-424. Available from: http://doi.wiley.com/10.3322/caac.21492

[82] Vecchione A, Baldassarre G, Ishii H, Nicoloso MS, Belletti B, Petrocca F, et al. Fez1/Lzts1 absence impairs Cdk1/Cdc25C interaction during mitosis and predisposes mice to cancer development. Cancer Cell [Internet]. 2007;**11**(3):275-289. Available from: https://linkinghub.elsevier.com/retrieve/pii/S153561080700030X

Chapter 6

Research Progress of DNA Methylation in Thyroid Cancer

Zhu Gaohong and Xie Lijun

Abstract

We have summarized increasing data from all kinds of experiment results of papers in recent years, which are associated with tumor suppressor genes, oncogenes, and thyroid-specific genes and attempt to elucidate the importance of epigenetic modifications and the mechanisms of aberrant DNA methylation in thyroid cancer in this review. The results showed that current articles have revealed the importance of epigenetic modifications and the different types of mechanisms in thyroid cancer. The mechanisms of DNA methylation related to thyroid cancer demonstrate that acquired epigenetic abnormalities together with genetic changes play an important role in alteration of gene expression patterns. Aberrant DNA methylation has been well known in the CpG regions. Among the genes identified, we have shown the status of DNA promoter methylation in papillary, follicular, medullary, and anaplastic thyroid cancer. It suggested that thyroid cancer subtypes present differential promoter methylation signatures, which will encourage potential thyroid cancer detection in its early stages, assessment of prognosis, and targeted cancer treatment.

Keywords: thyroid carcinoma, DNA methylation, epigenetic inheritance, tumor suppressor genes, oncogene genes, thyroid-specific genes

1. Introduction

Thyroid cancer is the most frequent endocrine neoplasia. The National Cancer Institute estimated that there would be 44,670 new cases of thyroid cancer (TC) with 1690 deaths in 2010, and with an overall estimate of 56,870 new cases by 2017, and its incidence has been increasing in recent decades. Compared with other adult cancers, TC tends to occur in younger people between the ages of 20 and 60. It is three times more common in women than men [1] and has the fastest rising incidence rates in women and the second fastest in men with an annual percentage change of approximately 5%, making TC the sixth most common cancer in women [2]. There are four main types of which papillary and follicular (PTC, FTC) types together account for >90% followed by medullary thyroid cancers (MTC) with 3–5% and anaplastic carcinomas (ATC) making up <3% [3]. Reasons for this trend have been attributed to improvement in imaging (ultrasound technology) that is allowing the identification of ever smaller thyroid nodules. However, with this gain in detection, determining which benign nodules (adenomas) will progress to cancer cannot be determined on the basis of histology alone, underscoring the need for genetic markers of early detection for TC. Recently, epigenetic alterations have been shown to play a role in the development and progression of thyroid cancer.

With the deepening of tumor research, it has been gradually found that epigenetics plays an important role in the occurrence and development of tumors. Mechanisms of epigenetics include, but are not limited to, DNA methylation (methylating of ciliary carbon at position 5), posttranslation modification of histone, chromatin remodeling (structural change), gene imprinting, RNA interference (noncoding RNA or gene silencing), etc. The epigenetic mechanisms of tumor cells have lost a fine regulation, and the breakdown of epigenetic patterns will lead to tumor phenotype expression. These mechanisms have been reviewed elsewhere, and here we will focus on DNA methylation in thyroid cancer.

DNA methylation is an important epigenetic change, which is persistent and hereditary. The methylation of promoter CpG can regulate gene expression and maintain chromosome integrity and DNA recombination. Based on 789 samples from the Gene Expression Omnibus (GEO) and the Cancer Genome Atlas (TCGA) databases, the five-CpG signature could provide a novel biomarker with useful applications in thyroid cancer (PTC, FTC, ATC, and MTC) diagnosis and the diagnostic score formula on the condition of DNA methylation data [4]. However, the methylation pattern is unstable and can be reversed by small molecules and endogenous enzymes, leading to dedifferentiation and tumor heterogeneity. Abnormal DNA methylation, including a decrease in the overall methylation level of the genome, was accompanied by hypermethylation in some gene promoter regions. The hypermethylation of tumor suppressor gene promoter can reduce its expression, while the hypomethylation of tumor suppressor gene promoter can increase its expression, leading to tumorigenesis [5–8]. The following will describe the state of aberrant DNA methylation in different thyroid cancers.

2. DNA methylation in PTC

RET/PTC rearrangement and mutations in Ras and BRAF genes often occur in papillary thyroid carcinoma (PTC) [9]. In addition, many methylation of cancer suppressor genes are associated with BRAF gene mutations, such as Ras-association domain family 1A (RASSF1A), solute carrier family 5 member 8 gene (SLC5A8), retinoic acid receptor β2 (RARβ2), tissue inhibitor of metalloproteinase3 (TIMP3), phosphatase and tensin homolog deleted on chromosome ten (PTEN), metallothionein 1G (MT1G), ataxia-telangiectasia mutated (ATM), E-cadherin (ECAD), death-associated protein kinase (DAPK), multiple tumor suppressor 1 (MTS1 or P16), and mut-L homolog 1 (MLH1). Mutations of TSHR gene are not common in thyroid cancer, but high methylation and low expression of TSHR gene often occur. Recent studies have found that thyroid-specific genes (thyroid-stimulating hormone receptor and sodium/iodide symporter (TSHR and NIS), thyroid transcription factor-1 (TTF-1)) play an important role in occurrence and development of PTC. This part summarizes the related research on methylation genes in PTC in recent years (see **Table 1**).

2.1 DNA methylation of cancer suppressor genes in PTC

2.1.1 Ras association domain family 1 (RASSF1A)

RASSF1A is a member of Ras superfamily, which is located at chromosome 3 (exactly on 3p21.3). Hypermethylation of CpG islands in the RASSF1A promoter region contributes to epigenetic inactivation. It is a tumor suppressor gene widely

Sample sources	Main method	Related gene	Methylation (high/low)	Related gene expression (high/low)	BRAF mutation (+/−)	Cell pathway	Function	References
Tumor suppressor genes								
Patient tissue, normal people tissue	Western blotting analyses, RT-QPCR, fluorescent analysis	RASSF1A	High	Low	+	MAPK	Stabilize the microtubules	[12, 13]
		SLC5A8	High	Low	+		Sodium transporter	[17]
		RARβ2	High	Low	+		Negative regulation of cell cycle	[20]
		TIMP3	High	Low	+		Inhibitor of metalloproteinase	[23]
Patient tissue, normal people tissue	Methylation-specific PCR, RT-QPCR	PTEN	High	Low	+	PI3K/Akt	Inhibit PI3K/Akt pathway	[26]
		MT1G	High	Low	+		Metallothionein	[28]
		ATM	High	Low	+		Regulate cell cycle	[29]
Patient tissue, normal people tissue	Database query and retrospective medical chart, RT-PCR	ECAD	High	Low	+	Wnt/β-catenin	Mediate the adhesion of cells	[8]
Thyroid-specific genes								
Patient tissue, normal people tissue	Western blotting analyses, RT-QPCR	NIS	High	Low	+	TSHR/cAMP	Sodium transporter	[36]
		TSHR	High	Low	+		Thyrotropin receptor	[37]
		TTF-1	High	Low	+		Inhibit the pro-adipogenic response to pioglitazone	[40]
Other potential genes								
Patient tissue, thyroid cancer cell line	Western blotting analyses, RT-QPCR, bisulfite sequencing, and methylation-specific PCR	RASSF10	High	–	–	Others	Inhibit cell proliferation	[42]
		14-3-3σ	High	–	–		Cell cycle regulation	[45]
		RIZ1	High	–	–		Epigenetic mechanisms	[44]
		DACT2	High	–	–		Inhibiting Wnt signaling	[45]

Table 1.
Aberrant methylated genes in PTC.

expressed in various normal organs but is often deleted in tumors. It is speculated that BRAFV600E gene mutation in PTC regulated the RASSF1A-MST1-FoxO3 signaling pathway, which led to RASSF1A hypomethylation and affected the malignant degree of thyroid cancer. It is found that the methylation rate of RASSF1A in thyroid cancer is 15–75% [10]. Studies indicated RASSF1A methylation differed in PTC compared with normal thyroid and was correlated with extracapsular invasion inversely. It suggested that RASSF1A has a potential role as a molecular marker for characterization of PTC histopathology [11–15]. It is shown that hypermethylation of RASSF1A promoter region is 20–32% in PTC. Hypermethylation of RASSF1A in PTC was related to the multifocal and extracapsular invasion of tumors [16].

2.1.2 Solute carrier gene family 5A, member 8 (SLC5A8)

SLC5A8 is a passive iodine transporter located in the parietal membrane of thyroid follicular cells. SLC5A8 is not regulated by thyrotropin in normal thyroid tissues but methylated in thyroid tumors. Hypermethylation often occurred in the first exon of CpG islands in SLC5A8, which results in gene silencing and restoring expression inhibiting cancer cell growth. It has been pointed out that SLC5A8 was an anti-oncogene of colon cancer [17]. SLC5A8 was also frequently hypermethylated in thyroid cancer. Its function is unclear nowadays, but its hypermethylation might play a key role in the occurrence of thyroid cancer [18]. Studies revealed that SLC5A8 gene was highly methylated in typical PTC (90%) and only 20% in other types of PTC. In addition, low expression of SLC5A8 was also associated with BRAF T1796A, suggesting that SLC5A8 methylation may be important in MAPK pathway [19].

2.1.3 Retinoic acid receptor beta2

RARβ2 is a type of nuclear receptor that is activated by both all-trans retinoic acid and 9-cis retinoic acid, which has been shown to function as a tumor suppressor gene in different types of human tumors. It has been found that RARβ2 expression was decreased or deleted in tumors. It meant that RARβ2 inactivation was related to tumorigenesis. In the treatment of metastasis and recurrence of thyroid cancer, retinoic acid therapy could restore the iodine uptake ability of metastasis and then improved the efficacy of 131-I radiotherapy. Researches indicated that the methylation rate of RARβ2 in thyroid cancer was 14%, higher than that in normal thyroid tissues (7%). RARβ2 gene methylation was associated with BRAF gene mutation in Wnt/beta catenin pathway [20]. Studies found there was RARβ2 gene hypermethylation in thyroid cancer cell lines. And after treatment with 5-azacytidine, RARβ2 expression was significantly increased, and the growth of tumors was inhibited, while the inhibition still existed after removing 5-azacytidine. RARβ2 gene methylation took part in tumorigenesis and development in PTC [21].

2.1.4 Tissue inhibitor of metalloproteinases-3 (TIMP3)

TIMP3 can bind to matrix metalloproteinases (MMPs), inhibiting the activity of MMPs effectively [20, 22]. Methylation of TIMP3 promoter has been demonstrated in many malignant tumors. It is often associated with growth, invasion, and lymph node metastasis of malignant tumors. It is pointed out that BRAF mutation caused low expression of TIMP3 in PTC, which could cause invasion and progression of tumors. It was found that 38% of TIMP3 are hypermethylated in PTC [23].

2.1.5 Phosphatase and tensin homolog deleted on chromosome 10 (PTEN)

The PTEN gene is located on chromosome 10 (especially on the region of 10q23), which could encode a specific phosphatidylinositol triphosphate 3 dephosphorylation and inhibit the activation of PI3K/Akt signaling pathway. PTEN expression alteration is crucial to the pathogenesis of cancer and other diseases. Low level of PTEN caused by homozygous deletions, frameshift, nonsense mutations or hypermethylation, or PTEN protein destability occurs frequently in various human cancers [24, 25]. It was shown that PTEN gene expression was low in thyroid cancer. One study found the methylation status of PTEN in FA, FTC, and PTC. The results indicated that PTEN methylation level was gradually increased in PTC (45.7%), FA (83.3%), and FTC (85.7%). PTEN methylation was related to mutations genes in PI3K/Akt signaling pathway, such as PIK3CA and Ras genes, suggesting that PTEN methylation and PI3K/Akt signaling pathway played an important role in the process of occurrence and development in PTC [26].

2.1.6 Metallothionein 1G (MT1G)

MT1G, a member of the metallothionein family, is a highly conserved cysteine-rich small molecule, which is mainly involved in metal-related transport. MT1G exists in normal cells; it can regulate and maintain intracellular metal ion balance, cell proliferation, and apoptosis. MT1G promoter methylation is associated with decreasing gene expression, but not complete abrogation. Studies have shown that MT1G gene has abnormal methylation in thyroid cancer, liver cancer, colon cancer, and prostate cancer. It is confirmed that restoring MT1G gene expression could inhibit tumors growth in vivo and in vitro, suggesting that MT1G gene has anticancer effect [27]. It is shown that MT1G gene was abnormally methylated in thyroid cancer (30.3% in malignant tumors and 18.8% in benign tumors). Its expression was significantly decreased, and that methylation of MT1G gene was associated with its low expression. Further studies suggested that restoring MT1G gene expression could inhibit the growth and infiltration of PTC and induced cell cycle inhibition and apoptosis. The mechanism may inhibit PI3K/AKT pathway. In addition, hypermethylation of MT1G was also associated with lymph node metastasis [28].

2.1.7 Ataxia-telangiectasia mutated

ATM belongs to the PI3/PI4 enzyme family. Although there is emerging evidence for a role of ATM in promoting tumorigenesis, ATM signaling provides a barrier to activated oncogenes and tumor progression, rather than promoting cancer early in tumorigenesis. ATM is ubiquitous in human and other higher animal tissues and cells, such as testicular tissue. And a study showed that ATM was hypermethylated in PTC (50%) and 0% in normal thyroid tissue [29].

2.1.8 E-cadherin

ECAD is a family of transmembrane glycoproteins responsible for calcium-dependent cell adhesion. It is the key structural components of adherens junctions. It is reported that ECAD may also act as a gene transcriptional regulator. Two main mechanisms including hypermethylation of the promoter and microRNA imbalance have been widely studied under the ECAD regulation in head and neck tumors. The methylation of ECAD promoter region was accounted for 39.3% in PTC, and ECAD expression decreased in the early stage of tumorigenesis. The experiment data showed that methylation level of ECAD in thyroid cancer increased to 56% (18/32) and 0% (0/27) in normal thyroid tissue. Further studies found that there was no significant correlation between ECAD methylation and T stage of lymph node metastasis in thyroid cancer. After 2.6 years of follow-up, the recurrence of thyroid cancer associated with ECAD methylation has no correlation [8].

2.1.9 Death-associated protein kinase

DAPK is a calmodulin-regulated ATK, which has an important role in the process of apoptosis. DAPK mechanism is largely due to promoter hypermethylation, leading to gene silencing. DAPK is ubiquitous in normal tissues. When the promoter of DAPK is methylated, it will cause an abnormal gene expression. Abnormal expression of DAPK can hinder the normal process of apoptosis and bring about tumorigenesis. And its low expression or deletion is one of the important mechanisms of cell carcinogenesis. It has been found that methylation of CpG island in the promoter region in DAPK is an important reason for expression silencing. The loss of DAPK protein in sporadic colorectal cancer is caused by the promoter hypermethylation. It existed in very small tumors. Therefore, the loss of DAPK gene plays an important role in the early stage of tumor formation. It is reported that DAPK promoter methylation accounts for 51% in PTC and abnormal methylation and DAPK gene silencing existed in many kinds of cancer cells [30]. It is pointed out that the high methylation level of DAPK gene was associated with tumor size and multiple lesions [31].

2.1.10 Multiple tumor suppressor 1 (MTS1 or P16)

MTS1 is hereafter called p16. It is an anti-oncogene in many tumors. The 5′-CpG fragment in the promoter region of P16 gene is the most susceptible to methylation, inhibiting its expression products. Abnormal expression of P16 gene can over-activate cyclin-dependent protein kinase 4 and stimulate abnormal cell proliferation, leading to tumorigenesis. Aberrant methylation of promoter region in P16 gene is the main cause for P16 gene inactivation [32]. Some scholars reported that the P16 gene was hypermethylated (35.9%) in PTC [33].

2.1.11 Mut-L homolog 1

MLH1 is one of the DNA mismatch repair genes located on the 3p21 region in chromosome 3. A correlation between MLH1 promoter methylation, specifically the 'C' region stops in MLH1 protein formation, can prevent the normal activation of DNA repair gene. Low expression of MLH1 gene was associated with BRAFV600E mutation and RET/PTC rearrangement. Hypermethylation of MLH1 promoter was found in colon cancer as well. A study indicated that abnormal methylation of MLH1 was significantly correlated with lymph node metastasis of PTC, suggesting that MLH1 might be a molecular marker of lymph node metastasis in PTC [34]. Another study found that there were abnormal methylation and low expression of MLH1 in thyroid cancer and MLH1 expression is associated with BRAF, IDH1, and NRAS gene mutations [35].

2.2 DNA methylation of thyroid-specific genes in PTC

2.2.1 Thyroid-stimulating hormone receptor and sodium/iodide symporter

On the other hand, methylation of thyroid-specific genes is also one of the causes for occurrence and development of PTC. Thyroid-specific genes mainly include TSHR, NIS, thyroglobulin (Tg), and thyroid peroxidase (TPO), which participate in thyroid iodine uptake and maintenance of normal thyroid function. Under normal conditions, TSH stimulates TSHR on thyroid follicular epithelial cells and activates NIS to ingest iodine into cells. TSH is produced by the pituitary thyrotrophs and stimulates thyroid functions using TSHR. The iodine ingested synthesizes thyroid hormones catalyzed by TPO and is stored in Tg. It is found that abnormal expression of these molecules is related to iodine metabolism in PTC. Studies have shown that the expression of TSHR, NIS, Tg, and TPO with BRAF mutated in thyroid cancer is decreased [36]. In PTC, both TSHR and NIS are abnormally methylated, and their expression is decreased. Low expression of TSHR and NIS may be related to the occurrence and development of tumors. It also reduces the uptake of iodine capacity in tumor cells. Scholars found TSHR and NIS become an important cause for PTC in 131-I radiotherapy [37, 38]. In human and rabbit thyroid cancer cells, BRAFV600E mutant, a carcinogenic homolog of murine sarcomatous virulent bacterium, could cause activation of BRAF/MEK/MAPK signaling pathway and expression silencing of thyroid-specific genes including TPO, Tg, TSHR, and NIS. At last, it resulted in the reduction of iodine uptake in PTC.

2.2.2 Thyroid transcription factor-1

TTF-1 is known as thyroid-specific enhancer-binding protein (T/EBP). It is a transcription factor with homologous domains in the thyroid, lung, and central nervous system. TTF-1 gene is located in region 14q13.3 on chromosome 14. It is comprised of three exons and two introns. Under physiological conditions, TTF-1 is stable positive in thyroid tissue. TTF-1 can regulate the expression of thyroid-related genes such as TG, TPO, TSHR, and NIS. Thus it acts a pivotal part in regulating growth, development, and function of thyroid. It showed hNIS mRNA expression loss might be related to methylation of thyroid-specific transcription factor genes. Abnormal methylation caused loss of transcription factor expression with indirect loss of hNIS mRNA expression through the KAT-5 and KAT-10 responses to 5-azacytidine treatment with acquisition of parallel TTF-1 and hNIS mRNA expression. It was found that insufficient expression of TTF-1 and Pax-8 may result in the decrease of activity of thyroglobulin gene promoter in thyroid cancer cells. Some confirmed that TTF-1 gene was expressed lowly in thyroid cancer [39]. Other researchers studied the methylation status of TTF-1 in thyroid cancer and found that TTF-1 gene was highly methylated and lowly expressed in thyroid cancer cell lines, but not in normal thyroid [40].

2.3 DNA methylation of oncogene gene in PTC

At present, BRAF gene mutation has been found in melanoma, ovarian serous tumor, colorectal cancer, glioma, liver cancer, and leukemia. A large number of studies also showed that BRAF gene mutation in PTC is closely related to methylation of tumor suppressor. BRAF gene is an important transducer for Ras/Raf/MEK/ERK/MAPK pathways. About 90% of T1799A point mutation in BRAFV600E happen in PTC, which can increase BRAF activity. The BRAF gene plays its biological role by activating MEK/ERK signaling pathway. The result showed that PTC with BRAF gene mutation had strong tissue invasiveness and was easy to infiltrate tissues around thyroid gland [41]. Methylation of TIMP3 gene, SLC5A8 gene, and DAPK gene, which are tumor suppressor genes, are related to BRAF gene mutation and PTC invasion. The overexpression of BRAFV600E gene could lead to silencing of some thyroid-specific genes (like NIS, TG, TPO) in the activation state of

BRAF/MEK/MAPK pathway. It could cause iodine uptake activity decrease and ineffectiveness of radioiodine therapy. Therefore, it could affect thyroid cancer progression.

2.4 Other potentially aberrant methylation genes in PTC

In recent years, DNA methylation has become a new research hotspot. Many genes have been studied as potential abnormal methylation sites, including Ras association domain family 2 (RASSF2), Ras-association domain family 10 (RASSF10), disheveled-binding antagonist of beta-catenin 2 (DACT2), retinoblastoma protein-interacting zinc finger gene 1 (RIZ1), 14-3-3 protein family (14-3-3 sigma), and other signaling pathways such as JAK-STAT pathway, NF-kappa B pathway, HIF1 alpha pathway, and Notch pathway in thyroid cancer [42–45]. However, there are relatively few reports on these genes and pathways in thyroid cancer. Most of the mechanisms are not clear at present and need further study.

3. DNA methylation in FTC

DNA methylation analysis revealed 2130 and 19 differentially methylated CpGs in PTC and follicular thyroid carcinoma (FTC), respectively [46]. Aberrant DNA methylation of tumor suppressor genes is common in FTC and ATC. Certain specific tumor suppressor genes are mainly PTEN, RASSF1A, Rap1-GTPase activating protein, and thyroid-specific gene TRSH in FTC. This part summarizes the related research on methylation genes in FTC in recent years (see **Table 2**).

References	Gene	Function	DNA methylation	BRAFE	Incidence%
Tumor suppressor genes					
[39–42]	RASSFIA	RASSF1A localizes to microtubules and promotes their stabilization	↑	+	75% of FTC
[50]	RAP1GAP	RAP1GAPase-activating protein	↑	+	38% of FTC
[25]	PTEN	PTEN is involved in the regulation of cell cycle and preventing cells from growing and dividing rapidly	↑	+	85.7% of FTC
Thyroid-specific genes					
[58]	NIS	Sodium transporter	↑	+	53.8% of thyroid cancers
[58]	TSHR	Thyrotropin receptor	↑	+	47% of FTC
Oncogene genes					
[52, 53]	Maspin	A member of serine protease inhibitor	↑	+	100% of WDTC
Other genes					
[61]	DMCpG	—	↑	—	84% of FTC
[61]	RASAL1	—	↑	—	4.88% of FTC
[57]	COL4A2	—	↑	—	56% of FTC
[64]	RASSF10	—	↑	—	50% of FTC

Table 2.
Aberrant methylated genes in FTC.

3.1 DNA methylation of tumor suppressor genes in FTC

3.1.1 PTEN

PTEN negatively regulates AKT/PKB signaling pathway. It is involved in regulation of cell cycle, cell growth inhibition, and rapid division [47]. Aberrant DNA methylation in this gene is also mostly reported in FTC. It confirmed that PTEN promoter hypermethylation was detected in six of seven (85%) FTC and five of six (83.3%) follicular adenomas. The results showed a high frequency of PTEN promoter hypermethylation, especially in follicular tumors. It means that it has a possible role in thyroid cancer [25]. Studies found methylation status of PTEN in FA and FTC. And PTEN methylation level was gradually increased in FA (83.3%) and FTC (85.7%). Authors have shown that methylation of PTEN promoter plays an important role in FTC [48].

3.1.2 RASSF1A

Differential expression of RASSF1A gene is related to occurrence of thyroid cancer. Aberrant DNA methylation is an important mechanism of RASSF1A gene inactivation. Studies found 44% benign adenomas, 75% follicular thyroid cancers tumors, and 20% PTC tumors harbored promoter methylation in greater than or equal to 25% of RASSF1A alleles by real-time quantitative methylation-specific PCR [39–42]. Methylation frequency was higher in invasive thyroid cancer. It was found to be 70% of the RASSF1A methylation rate in FTC, 80% in MTC, and 78% in UTC compared with benign PTC [14]. Inactivation of RASSF1A in different stages of thyroid cancer was detected by tumor metastasis classification, and compared with FTC, only a small part of RASSF1A methylation in PTC is abnormal. These studies indicated follicular cell-derived thyroid tumorigenesis may be an early step [15, 49].

3.1.3 Rap1-GTPase-activating protein

RAP1GAP gene encodes a type of GTPase-activating protein that downregulates Ras-related protein activity. Ras oncoproteins are very important for both development and maintenance of many tumor types. RAP1GAP is involved in the regulation of mitosis and carcinogenesis in thyroid cells. Researchers aimed to determine the global patterns of aberrant DNA methylation in thyroid cancer using DNA methylation arrays [50]. And the study identified 262 and 352 hypermethylated and 13 and 21 hypomethylated genes in PTC and FTC, respectively. In addition, 86 and 131 hypermethylated genes were identified. Among these genes, four potential oncogenes (INSL4, DPPA2, TCL1B, and NOTCH4) were frequently regulated by aberrant methylation in primary thyroid tumors [51].

3.1.4 Mammary serine protease inhibitor (Maspin)

Besides, a member of the serine protease inhibitor superfamily named Maspin is a unique tumor suppressor gene encoding SERPINB5 gene. Epigenetic changes of Maspin expression occurred in the 5′ regulatory region of Maspin gene and involved cytosine methylation, histone deacetylation, and chromatin accessibility. The epigenetic deregulation frequently participates in tumorigenesis by inactivation of tumor suppressor genes. The association of promoter hypermethylation and gene silencing is an established oncogenic process in cancer. Promoter methylation of Maspin gene could lead to gene silencing in thyroid cancer, breast cancer, skin cancer, and colon cancer. Studies have indicated that overexpression of Maspin in gastric cancer, pancreatic cancer, and ovarian cancers resulted from CpG promoter of Maspin demethylation. A study detected DNA methylation status in Maspin promoter region, indicating that overexpression of the gene was the result of DNA hypomethylation [52]. It was closely related to the morphological dedifferentiation of thyroid cancer. Another study found 100% Maspin hypermethylation was closely associated with morphological dedifferentiation in thyroid cancers [52, 53].

3.2 DNA methylation of thyroid-specific genes in FTC

In addition to tumor suppressor genes and oncogenes, hypermethylation could lead to NIS and TSHR gene silencing in FTC, too. NIS methylation is of great significance in treatment of thyroid cancer. Therefore, abnormal methylation of these genes may be the pathogenesis or progression factor in FTC [54, 55]. Aberrant promoter methylation was examined in 24 tumor suppressor genes using methylation-specific multiplex ligation-dependent probe amplification (MS-MLPA) assay and methylation-specific PCR (MSP). In thyroid cancers, CASP8 (caspase-8), RASSF1, and NIS were methylated in 9/13, 10/13, and 7/13, respectively [56]. Some researches also found combination gene panels TPO and UCHL1 (ROC = 0.607, sensitivity 78%) discriminated FTC from FA and RASSF1 and TPO (ROC = 0.881, sensitivity 78%) discriminated FTC from normal. Methylation of TSHR distinguished PTC from FTC (ROC = 0.701, sensitivity 84%) and PTC from FA (ROC = 0.685, sensitivity 70%) [57]. And the six-gene panel of TIMP3, RARβ2, SERPINB5, RASSF1, TPO, and TSHR, which differentiates PTC from normal thyroid, had the best combination sensitivity (91%) and specificity (81%) of the panels addressing discrimination of cancer tissue by quantitative methylation-specific polymerase chain reaction (QMSP) in a retrospective cohort of 329 patients [58, 59].

3.3 Other abnormal methylation genes in FTC

Nowadays, there are a few studies on methylation of FTC genes. However, many genes still have been studied as potential abnormal methylation sites in FTC. For example, one of the ZIC families called ZIC1 (C2H2-type zinc finger proteins) is frequently hypermethylated in FTCs [60]. Moreover, 3564 differentially methylated CpGs (DMCpG) were detected in FTC and 84% hypermethylated with respect to normal controls. It is suggested that perturbed DNA methylation, in particular hypermethylation, is a component of the molecular mechanisms leading to FTC formation and that DNA methylation profiling might help in differentiating FTCs from their benign counterpart [61]. Also, others identified the presence of RASAL1 mutations, with a prevalence of 4.88% (n = 2 of 41) in FTC and 16.67% (n = 5 of 30) in ATC [62]. Studies found a more detailed analysis showing that 53.9% of the hypermethylated and 81.5% of the hypomethylated CpG sites identified in differentiated primary tumors (PTCs and FTCs) were also present in differentiated thyroid carcinoma-derived cancer cell [61]. Aside from that, COL4A2 was hypermethylated in 56% of the FTC samples by array measurement in the discovery series [57, 63]. Another study indicated that RASSF10 was frequently hypermethylated in thyroid cancer. It showed 50% of methylation frequency of RASSF10 in FTC and the highest (100%) in MTCs [64].

4. DNA methylation in ATC

Promoter methylation of PTEN is also common in anaplastic thyroid cancer (ATC) [65]. PTEN methylation is related to gene changes of PI3K Akt pathway in thyroid tumors, including PTEN mutations, various subtypes of Ras mutations, PIK3CA mutations, and amplification [66, 67]. One study analyzed 24 genetic alterations in the major genes of MAPK and PI3K-AKT pathways in 48 ATC samples and found that the majority of (81%) samples that harbored genetic alterations could be likely activated in both pathways [68]. Accordingly, another DNA methylation pan-cancer study focused on promoters found that thyroid carcinoma exhibited one of the lowest frequencies in both hypomethylation and hypermethylation events. And ATC exhibits a high frequency of DNA methylation alterations (tenfold higher than PTC) [69, 70]. A recent pan-cancer analysis on whole exome sequencing revealed that the mutation frequency in PTC was one of the lowest (approximately 1 change/Mb across the entire exome) among solid tumors, while the mutation frequency in ATC was at the opposite extreme and was closer to that in melanoma and lung cancer, exceeding 100 changes/Mb [71, 72].

In addition, solute carrier family 26, member 4 (SLC26A4) gene, encodes a transmembrane protein named pendrin with up to 15 predicted membrane spanning domains and affects the flow of iodine into follicular lumen. The following were reported: 71% of ATC, 44% of benign tumors, 46% of FTC, and 71% of PTC, with abnormal SLC26A4 gene methylation in 64 cases of primary thyroid tumors and 6 cases of thyroid tumor cell lines [18, 73]. In addition, 81.5% of hypermethylated genes and 89% of hypomethylated genes were also present in nondifferentiated primary tumors (MTCs and ATCs) and nondifferentiated thyroid carcinoma-derived cancer cell lines [74], while Ras protein activator like-1 gene (RASAL1) displayed MAPK- and PI3K-suppressing and thyroid tumor-suppressing activities, which can be impaired by the mutations. Hypermethylation and mutations of RASAL1 were found in 33.33% (n = 10 of 30) of ATCs and in 0 of 20 (0%) of benign thyroid tumors [62]. However, ATC showed more hypomethylation than hypermethylation events, indicating that hypomethylation is related to dedifferentiation [70]. The authors validated four genes (NOTCH4 and TCL1B in ATCs, INSL4 and DPPA2 in MTCs) that become aberrantly hypomethylated in nondifferentiated thyroid tumors. All of them have been proposed to have an oncogenic role in cancer. And NOTCH4 (a member of the Notch family of transmembrane receptors) is frequently overexpressed in thyroid tumors [75, 76].

This part summarizes the related research on methylation genes in ATC in recent years (see **Table 3**).

Authors	Gene	Function	DNA methylation	BRAFE	Incidence%
Oncogene genes					
[70]	TCL1B	An oncogene frequently activated by reciprocal translocations	↑	+	64% of ATC
[70]	NOTCH4 A	Member of notch family, which plays a role in a variety of developmental processes	↑	+	45% of ATC
Thyroid-specific genes					
[18, 73]	SLC26A4	Dysfunctional pendrin	↑	—	71% ATC
Tumor suppressor genes					
[25]	PTEN	PI3K–AKT pathway	↑	—	81% of ATC
[62]	RASAL1	MAPK- and PI3K-suppressing	↑	—	33.33% of ATC

Table 3.
Aberrant methylated genes in ATC.

5. DNA methylation in MTC

Medullary thyroid cancer is a neuroendocrine tumor originating from parafollicular C cells, and it is highly resistant to chemo- and radiotherapy [77]. Spry1 is a candidate tumor-suppressor gene in MTC. The hyper-activation of PI3K/Akt/mTOR cascade has a relevant role in the pathogenesis and progression of MTC. In fact, most of pro-oncogenic effects of RET and Ras mutations are modulated by the activation of PI3K/Akt/mTOR pathway [78, 79]. For patients with metastatic disease, standard treatment modalities include local treatments (radiofrequency ablation, radiation therapy, embolization) and systemic treatment chemotherapy and more recently tyrosine kinase inhibitors (TKIs) targeting RET protein. As experience has been limited to case reports or case series, response rates (RR) ranged from 0 to 25% for periods of up to a few months. In nondifferentiated thyroid tumors, INSL4 and DPPA2 become aberrantly hypomethylated, both of which have been proposed to have an oncogenic role in MTC.

5.1 Sprouty1 (Spry1)

Sprouty (Spry) family of genes is composed of four members in mammals (Spry1–4). The Spry1 promoter is frequently methylated in MTC, and that Spry1 expression is consequently decreased. These findings identify Spry1 as a candidate tumor-suppressor gene in MTC. In mammals, the situation is more complicated as Spry proteins have been shown to be activators or inhibitors of receptor tyrosine kinase signaling depending on the cellular context or the receptor tyrosine kinase analyzed [80, 81]. Spry family members have been proposed to function as tumor-suppressor genes in a growing list of cancerous malignancies, including prostate and hepatocellular carcinoma, B-cell lymphoma, or neuroblastoma [82]. Finally, the authors found that SPRY1 promoter is frequently methylated and its expression decreased in human MTC. The mRNA levels of Spry1 are detected in murine C cells of thyroid with real-time RT-PCR. In situ hybridization showed expression of Spry1 mRNA in the fourth pharyngeal pouch, where thyroid C-cells originate and beta-galactosidase staining of thyroids from 1-month-old Spry1LacZ/þ mice. One study used the hypomethylating agent 5′-aza-deoxycytidine (5′-Aza-dC) to confirm a causal relationship between promoter methylation and Spry1 expression [83]. As expected, TT cells treated with 5′-aza-dC showed an increase of approximately sixfold in the levels of Spry1 mRNA when compared to vehicle-treated cells. 5′-Aza-deoxycytidine (AZA), a demethylating agent, is in combination with the mTOR inhibitor everolimus in MTC cells (MZ-CRC-1 and TT). An innovative bioinformatic pipeline identified four potential molecular pathways implicated in the synergy between AZA and everolimus: PI3K-Akt signaling, the neurotrophin pathway, ECM-receptor interaction, and focal adhesion. Among these, the neurotrophin signaling pathway was most directly involved in apoptosis, through NGFR and Bax gene overexpression. Increased expression of genes involved in the NGFR-MAPK10-TP53-Bax/Bcl2 pathway during incubation with AZA plus everolimus was validated by western blotting in MZ-CRC-1 cells [84].

5.2 Insulin-like 4 (INSL4)

INSL4 (pro-EPIL) belongs to the insulin and insulin-like growth factor family and is expressed strongly during the first trimester of pregnancy by the differentiated syncytiotrophoblast [85]. It has been shown to be overexpressed in breast tumors with an aggressive phenotype [86], but the underlying mechanisms are still unknown. The aberrant overexpression of INSL4 in breast tumors, together with the aberrant promoter hypomethylation reported in this study, suggests that promoter demethylation might be a frequent mechanism of activation of INSL4 oncogene activation in cancer.

5.3 Developmental pluripotency-associated 2 (DPPA2)

DPPA2 is expressed early in the embryo's development [87] but also in some tumor types [88]. Although the underlying molecular mechanism has not been reported yet, the authors' data indicated that promoter hypomethylation might play an important role. The frequent promoter hypomethylation observed in nondifferentiated tumors might be relevant for treatment with demethylating drugs [51, 88].

6. Conclusion

More and more researches have realized that the occurrence of tumors is not only entirely determined by genes but also epigenetics. The changes of epigenetics in thyroid cancer are mainly manifested in the aberrant methylation of tumor suppressor genes and thyroid-related genes. Numerous studies on DNA methylation in thyroid cancer have improved our understanding of thyroid carcinogenesis. Some of the recent findings, including the huge catalog of DNA methylation alterations, the association of DNA hypomethylation with cancer progression and dedifferentiation, the existence of different methylomes related to different clinical and molecular phenotypes, and the influence of immune-infiltrating cells in tumor DNA methylation patterns, are most likely to lead the direction of future research in the field of DNA methylation in thyroid cancer. A large number of studies confirmed the importance of DNA methylation as a source of novel biomarkers for early diagnosis, therapeutic perspective, and prognosis evaluation in thyroid cancer. In addition, the design of specific target demethylation drugs, which reactivate the function of tumor suppressor genes, is expected to become a new scheme for cancer treatment. Therefore, further functional experiments in vitro and in vivo are necessary for better understanding of the meaning and potential mechanism of DNA methylation changes in thyroid cancer as well as the evaluation of candidate biomarkers through case-control studies and prospective trials.

Author details

Zhu Gaohong* and Xie Lijun
Department of Nuclear Medicine, First Affiliated Hospital of Kunming Medical University, Kunming, China

*Address all correspondence to: 1026909611@qq.com

References

[1] Smith RA, Andrews KS, Brooks D, Fedewa SA, Manassaram-Baptiste D, Saslow D, et al. Cancer screening in the United States, 2018: A review of current American Cancer Society guidelines and current issues in cancer screening. CA: A Cancer Journal for Clinicians. 2018; **68**(4):297-316. DOI: 10.3322/caac.21446

[2] Stearns V, Yamauchi H, Hayes DF. Circulating tumor markers in breast cancer: Accepted utilities and novel prospects. Breast Cancer Research and Treatment. 1998;**52**(1–3):239-259. DOI: 10.1023/a:1006137619153

[3] Quaglino F, Marchese V, Lemini R, Piovesan A, Mazza E, Viora T, et al. Parathyroid carcinoma. A single Institution experience and a review of the international literature. Annali Italiani di Chirurgia. 2018;**89**:295-304

[4] Jia X, Zhao Q, Zhang Y, Dong Y, Lei L, Williamson RA, et al. Identification of a five-CpG signature with diagnostic value in thyroid cancer. Journal of Computational Biology. 2019; **26**(12):1409-1417. DOI: 10.1089/cmb.2019.0165

[5] D'Cruz AK, Vaish R, Vaidya A, Nixon IJ, Williams MD, Vander Poorten V, et al. Molecular markers in well-differentiated thyroid cancer. European Archives of Oto-Rhino-Laryngology. 2018;**275**(6):1375-1384. DOI: 10.1007/s00405-018-4944-1

[6] Ahmed AA, Essa MEA. Potential of epigenetic events in human thyroid cancer. Cancer Genetics. 2019;**239**: 13-21. DOI: 10.1016/j.cancergen.2019.08.006

[7] Moore LD, Le T, Fan G. DNA methylation and its basic function. Neuropsychopharmacology. 2013;**38**(1): 23-38. DOI: 10.1038/npp.2012.112

[8] Smith JA, Fan CY, Zou C, Bodenner D, Kokoska MS. Methylation status of genes in papillary thyroid carcinoma. Archives of Otolaryngology–Head & Neck Surgery. 2007;**133**(10): 1006-1011. DOI: 10.1001/archotol.133.10.1006

[9] Argyropoulou M, Veskoukis AS, Karanatsiou PM, Manolakelli A, Kostoglou-Athanassiou I, Vilaras G, et al. Low prevalence of TERT promoter, BRAF and RAS mutations in papillary thyroid cancer in the Greek population. Pathology & Oncology Research. 2018:10-25. DOI: 10.1007/s12253-018-0497-2

[10] Schagdarsurengin U, Richter AM, Hornung J, Lange C, Steinmann K, Dammann RH. Frequent epigenetic inactivation of RASSF2 in thyroid cancer and functional consequences. Molecular Cancer. 2010;**9**:264. DOI: 10.1186/1476-4598-9-264

[11] Jiang J-L, Tian G-L, Chen S-J, Xu L, Wang H-Q. Promoter methylation of p16 and RASSF1A genes may contribute to the risk of papillary thyroid cancer: A meta-analysis. Experimental and Therapeutic Medicine. 2015;**10**(4): 1549-1555. DOI: 10.3892/etm.2015.265

[12] Nakamura N, Carney JA, Jin L, Kajita S, Pallares J, Zhang H, et al. RASSF1A and NORE1A methylation and BRAFV600E mutations in thyroid tumors. Laboratory Investigation; A Journal of Technical Methods and Pathology. 2005;**85**(9):1065-1075. DOI: 10.1038/labinvest.3700306

[13] Pfeifer GP, Dammann R. Methylation of the tumor suppressor gene RASSF1A in human tumors. Biochemistry (Mosc). 2005;**70**(5):576-583. DOI: 10.1007/s10541-005-0151-y

[14] Schagdarsurengin U, Gimm O, Hoang-Vu C, Dralle H, Pfeifer GP, Dammann R. Frequent epigenetic silencing of the CpG island promoter of

RASSF1A in thyroid carcinoma. Cancer Research. 2002;**62**(13):3698-3701. DOI: 10.1046/j.1432-1033.2002.02944.x

[15] Xing M, Cohen Y, Mambo E, Tallini G, Udelsman R, Ladenson PW, et al. Early occurrence of RASSF1A hypermethylation and its mutual exclusion with BRAF mutation in thyroid tumorigenesis. Cancer Research. 2004;**64**(5):1664-1668. DOI: 10.1158/0008-5472.can-03-3242

[16] Lee SJ, Lee MH, Kim DW, Lee SE, Huang S, Ryu MJ, et al. Cross-regulation between oncogenic BRAF(V600E) kinase and the MST1 pathway in papillary thyroid carcinoma. PLoS One. 2011;**6**(1):e16180. DOI: 10.1371/journal.pone.0016180

[17] Gurav A, Sivaprakasam S, Bhutia YD, Boettger T, Singh N, Ganapathy V. Slc5a8, a Na^+-coupled high-affinity transporter for short-chain fatty acids, is a conditional tumour suppressor in colon that protects against colitis and colon cancer under low-fibre dietary conditions. The Biochemical Journal. 2015;**469**(2):267-278. DOI: 10.1042/BJ20150242

[18] Zane M, Agostini M, Enzo MV, Casal Ide E, Del Bianco P, Torresan F, et al. Circulating cell-free DNA, SLC5A8 and SLC26A4 hypermethylation, BRAF (V600E): A non-invasive tool panel for early detection of thyroid cancer. Biomedicine & Pharmacotherapy. 2013; **67**(8):723-730. DOI: 10.1016/j.biopha.2013.06.007

[19] Porra V, Ferraro-Peyret C, Durand C, Selmi-Ruby S, Giroud H, Berger-Dutrieux N, et al. Silencing of the tumor suppressor gene SLC5A8 is associated with BRAF mutations in classical papillary thyroid carcinomas. The Journal of Clinical Endocrinology and Metabolism. 2005;**90**(5):3028-3035. DOI: 10.1210/jc.2004-1394

[20] Brait M, Loyo M, Rosenbaum E, Ostrow KL, Markova A, Papagerakis S, et al. Correlation between BRAF mutation and promoter methylation of TIMP3, RARbeta2 and RASSF1A in thyroid cancer. Epigenetics. 2012;**7**(7): 710-719. DOI: 10.4161/epi.20524

[21] Miasaki FY, Vivaldi A, Ciampi R, Agate L, Collecchi P, Capodanno A, et al. Retinoic acid receptor beta2 re-expression and growth inhibition in thyroid carcinoma cell lines after 5-aza-2′-deoxycytidine treatment. Journal of Endocrinological Investigation. 2008; **31**(8):724-730. DOI: 10.1007/BF03346422

[22] Christensen DRG, Brown FE, Cree AJ, Ratnayaka JA, Lotery AJ. Sorsby fundus dystrophy—A review of pathology and disease mechanisms. Experimental Eye Research. 2017;**165**: 35-46. DOI: 10.1016/j.exer.2017.08.014

[23] Anania MC, Sensi M, Radaelli E, Miranda C, Vizioli MG, Pagliardini S, et al. TIMP3 regulates migration, invasion and in vivo tumorigenicity of thyroid tumor cells. Oncogene. 2011;**30**(27): 3011-3023. DOI: 10.1038/onc.2011.18

[24] Cretella D, Digiacomo G, Giovannetti E, Cavazzoni A. PTEN alterations as a potential mechanism for tumor cell escape from PD-1/PD-L1 inhibition. Cancers (Basel). 2019;**11**(9). DOI: 10.3390/cancers11091318

[25] Alvarez-Nuñez F, Bussaglia E, Mauricio D, Ybarra J, Vilar M, Lerma E, et al. PTEN promoter methylation in sporadic thyroid carcinomas. Thyroid. 2006;**16**(1):17-23. DOI: 10.1089/thy.2006.16.17

[26] Wang L, Zhang T, Guo L, Lin H. Regulation of PTEN expression by noncoding RNAs. Journal of Experimental & Clinical Cancer Research. 2018;**37**:223. DOI: 10.1186/s13046-018-0898-9

[27] Zeng J-d, Zhang N, Zhao G-j, Xu L-x, Yang Y, Xu X-y, et al. MT1G is

silenced by DNA methylation and contributes to the pathogenesis of hepatocellular carcinoma. Journal of Cancer. 2018;**9**(16):2807-2816. DOI: 10.7150/jca.25680

[28] Jiao F, Lv H, Guan H, Ma X, Ji M, He N, et al. Metallothionein 1G functions as a tumor suppressor in thyroid cancer through modulating the PI3K/Akt signaling pathway. BMC Cancer. 2013;**13**(1):462-474. DOI: 10.1186/1471-2407-13-462

[29] Yulu G, Liu X, Yu Y, Shi J, Ai L, Sun H, et al. Association of ATM gene polymorphism with PTC metastasis in female patients. Int. The Journal of Endocrinology. 2014;**370825**. DOI: 10.1155/2014/370825

[30] Bai J, Xue Z, Hu K, Liu B, Wang H, Li A, et al. Silencing DNA methyltransferase 1 (DNMT1) inhibits proliferation, metastasis and invasion in ESCC by suppressing methylation of RASSF1A and DAPK. Oncotarget. 2016; **7**(28):44129-44141. DOI: 10.18632/oncotarget.9866

[31] Hu S, Ewertz M, Tufano RP, Brait M, Carvalho AL, Liu D, et al. Detection of serum deoxyribonucleic acid methylation markers: A novel diagnostic tool for thyroid cancer. The Journal of Clinical Endocrinology and Metabolism. 2006; **91**(1):98-104. DOI: 10.1210/jc.2005-1810

[32] Yang D, Zhang H, Hu X, Xin S, Duan Z. A bnormality of pl6/p38MAPK/p53/Wipl pathway in papillary thyroid cancer. Gland Surgery. 2012;**1**(1):33-38. DOI: 10.3978/j.issn.2227-684X.2012.04.01

[33] Ishida E, Nakamura M, Shimada K, Higuchi T, Takatsu K, Yane K, et al. DNA hypermethylation status of multiple genes in papillary thyroid carcinomas. Pathobiology. 2007;**74**(6): 344-352. DOI: 10.1159/000110028

[34] Guan H, Ji M, Hou P, Liu Z, Wang C, Shan Z, et al. Hypermethylation of the DNA mismatch repair gene hMLH1 and its association with lymph node metastasis and T1799A BRAF mutation in patients with papillary thyroid cancer. Cancer. 2008;**113**(2):247-255. DOI: 10.1002/cncr.23548

[35] Santos JC, Bastos AU, Cerutti JM, Ribeiro ML. Correlation of MLH1 and MGMT expression and promoter methylation with genomic instability in patients with thyroid carcinoma. BMC Cancer. 2013;**13**:79. DOI: 10.1186/1471-2407-13-79

[36] Li J, Dong JN, Zhao Z, Lv Q, Yun B, Liu JQ, et al. Expression of sodium/iodide transporters and thyroid stimulating hormone receptors in thyroid cancer patients and its correlation with iodine nutrition status and pathology. European Review for Medical and Pharmacological Sciences. 2018;**22**(14):4573-4580. DOI: 10.26355/eurrev_201807_15513

[37] Gunin AG, Golubtsova NN, Kravtsova OA, Subbotkin AS, Subbotkina NO, Filippov FN. Number, proliferative activity, and expression of thyroid hormone receptors in dermal fibroblasts in mice with changed thyroid status. Bulletin of Experimental Biology and Medicine. 2019;**166**(6):797-801. DOI: 10.1007/s10517-019-04443-x

[38] Song J, Qiu W, Deng X, Qiu Z, Fan Y, Yang Z. A somatic mutation of RasGRP3 decreases Na(+)/I(−) symporter expression in metastases of radioactive iodine-refractory thyroid cancer by stimulating the Akt signaling pathway. American Journal of Cancer Research. 2018;**8**(9):1847-1855. Available from: www.ajcr.us/ISSN: 2156-6976/ajcr0083696

[39] Xu B, O'Donnell M, O'Donnell J, Yu J, Zhang Y, Sartor MA, et al. Adipogenic differentiation of thyroid cancer cells through the Pax8-PPARgamma fusion protein is regulated

by thyroid transcription factor 1 (TTF-1). The Journal of Biological Chemistry. 2016;**291**(37):19274-19286. DOI: 10.1074/jbc.M116.740324

[40] Kondo T, Nakazawa T, Ma D, Niu D, Mochizuki K, Kawasaki T, et al. Epigenetic silencing of TTF-1/NKX2-1 through DNA hypermethylation and histone H3 modulation in thyroid carcinomas. Laboratory Investigation. 2009;**89**(7):791-799. DOI: 10.1038/labinvest.2009.50

[41] Argyropoulou M, Veskoukis AS, Karanatsiou PM, Manolakelli A, Kostoglou-Athanassiou I, Vilaras G, et al. Low prevalence of TERT promoter, BRAF and RAS mutations in papillary thyroid cancer in the Greek population. Pathology & Oncology Research. 2018. DOI: 10.1007/s12253-018-0497-2

[42] Cheng F, Wang W, Jin J, Yu Z, Xin X. RASSF10 is epigenetically inactivated and suppresses cell proliferation and induces cell apoptosis by activating the p53 signalling pathway in papillary thyroid carcinoma cancer. Cellular Physiology and Biochemistry. 2017;**41**(3):1229-1239. DOI: 10.1159/000464386

[43] Lala G, Padmanabha L, Provenzano M, Fitzgerald M, Weydert J, Domann FE. Regulation of 14-3-3sigma expression in human thyroid carcinoma is epigenetically regulated by aberrant cytosine methylation. Cancer Letters. 2008; **267**(1):165-174. DOI: 10.1016/j.canlet.2008.03.017

[44] Lal G, Padmanabha L, Smith BJ, Nicholson RM, Howe JR, O'Dorisio MS, et al. RIZ1 is epigenetically inactivated by promoter hypermethylation in thyroid carcinoma. Cancer. 2006; **107**(12):2752-2759. DOI: 10.1002/cncr.22325

[45] Zhao Z, Herman JG, Brock MV, Sheng J, Zhang M, Liu B, et al. Methylation of DACT2 promotes papillary thyroid cancer metastasis by activating Wnt signaling. PLoS One. 2014;**9**(11):e112336. DOI: 10.1371/journal.pone.0112336

[46] Barros-Filho MC, Dos Reis MB, Beltrami CM, de Mello JBH, Marchi FA, Kuasne H, et al. DNA methylation-based method to differentiate malignant from benign thyroid lesions. Thyroid. 2019;**29**(9):1244-1254. DOI: 10.1089/thy.2018.0458

[47] Arico S, Petiot A, Bauvy C, Dubbelhuis PF, Meijer AJ, Codogno P, et al. The tumor suppressor PTEN positively regulates macroautophagy by inhibiting the phosphatidylinositol 3-kinase/protein kinase B pathway. The Journal of Biological Chemistry. 2001; **276**(38):35243-35246. DOI: 10.1074/jbc.C100319200

[48] Pringle DR, Vasko VV, Yu L, Manchanda PK, Lee AA, Zhang X, et al. Follicular thyroid cancers demonstrate dual activation of PKA and mTOR as modeled by thyroid-specific deletion of Prkar1a and Pten in mice. Journal of Clinical Endocrinology and Metabolism. 2014;**99**(5):E804-E812. DOI: 10.1210/jc.2013-3101

[49] Brown TC, Juhlin CC, Healy JM, Prasad ML, Korah R, Carling T. Frequent silencing of RASSF1A via promoter methylation in follicular thyroid hyperplasia: A potential early epigenetic susceptibility event in thyroid carcinogenesis. JAMA Surgery. 2014;**149**(11):1146-1152. DOI: 10.1001/jamasurg.2014.1694

[50] Zuo H, Gandhi M, Martin M, Edreira DH, Nimgaonkar VL, Zhang P, et al. Downregulation of Rap1GAP through epigenetic silencing and loss of heterozygosity promotes invasion and progression of thyroid tumors. Cancer Research. 2010;**70**(4): 1389-1397. DOI: 10.1158/0008-5472.CAN-09-2812

[51] Rodríguez-Rodero S, Fernández AF, Fernández-Morera JL, Castro-Santos P, Bayon GF, Ferrero C, et al. DNA methylation signatures identify biologically distinct thyroid cancer subtypes. Journal of Clinical Endocrinology and Metabolism. 2013; **98**(7):2811-2821. DOI: 10.1210/jc.2012-3566

[52] Ogasawara S, Maesawa C, Yamamoto M, Akiyama Y, Wada K, Fujisawa K, et al. Disruption of cell-type-specific methylation at the Maspin gene promoter is frequently involved in undifferentiated thyroid cancers. Oncogene. 2004;**23**(5):1117-1124. DOI: 10.1038/sj.onc.1207211

[53] Boltze C, Schneider-Stock R, Quednow C, Hinze R, Mawrin C, Hribaschek A, et al. Silencing of the maspin gene by promoter hypermethylation in thyroid cancer. International Journal of Molecular Medicine. 2003;**12**(4):479-484. DOI: 10.3892/ijmm.12.4.479

[54] Kleiman DA, Buitrago D, Crowley MJ, BS TB, Veach AJ, Zanzonico MJ, et al. Thyroid stimulating hormone increases iodine uptake by thyroid cancer cells during BRAF silencing. Journal of Surgical Research. 2013;**1**(182):85-93. DOI: 10.1016/j.jss.2012.08.053

[55] Eze OP, Starker LF, Carling T. The role of epigenetic alterations in papillary thyroid carcinogenesis. Journal of Thyroid Research. 2011;**895470**. DOI: 10.4061/2011/895470

[56] Stephen JK, Chitale D, Narra V, Chen KM, Sawhney R, Worsham MJ. DNA methylation in thyroid tumorigenesis. Cancers (Basel). 2011; **3**(2):1732-1743. DOI: 10.3390/cancers3021732

[57] Mancikova V, Buj R, Castelblanco E, Inglada-Perez L, Diez A, de Cubas AA, et al. DNA methylation profiling of well-differentiated thyroid cancer uncovers markers of recurrence free survival. International Journal of Cancer. 2014;**135**(3):598-610. DOI: 10.1002/ijc.28703

[58] Xing M, Usade H, Cohen Y, Tokumaru Y, Guo Z, Westra WB, et al. Methylation of the thyroid-stimulating hormone receptor gene in epithelial thyroid tumors: A marker of malignancy and a cause of gene silencing. Cancer Research. 2003;**63**(9):2316-2321. DOI: 10.1097/00130404-200305000-00010

[59] Demura M, Yoneda T, Wang F, Zen Y, Karashima S, Zhu A, et al. Ectopic production of parathyroid hormone in a patient with sporadic medullary thyroid cancer. Endocrine Journal. 2010;**57**(2):161-170. DOI: 10.1507/endocrj.k09e-131

[60] Gan L, Chen S, Zhong J, Wang X, Lam EKY, Liu X, et al. ZIC1 is downregulated through promoter hypermethylation, and functions as a tumor suppressor gene in colorectal cancer. PLoS One. 2011;**6**(2):e16916. DOI: 10.1371/journal.pone.0016916

[61] Affinito O, Salerno P, D'Alessio AM, Cuomo M, Florio E, Carlomagno F, et al. Association between DNA methylation profile and malignancy in follicular-patterned thyroid neoplasms. Endocrine-Related Cancer. 2019;**26**(4): 451-462. DOI: 10.1530/ERC-18-0308

[62] Liu D, Yang C, Bojdani E, Murugan AK, Xing M. Identification of RASAL1 as a major tumor suppressor gene in thyroid cancer. Journal of the National Cancer Institute. 2013; **105**(21):1617-1627. DOI: 10.1093/jnci/djt249

[63] Geers C, Colin IM, Gérard AC. Delta-like 4/Notch pathway is differentially regulated in benign and malignant thyroid tissues. Thyroid. 2011;**21**(12):1323-1330. DOI: 10.1089/thy.2010.0444

[64] Schagdarsurengin U, Richter AM, Wöhler C, Dammann RH. Frequent epigenetic inactivation of RASSF10 in thyroid cancer. Epigenetics. 2009;**4**(8): 571-576. DOI: 10.4161/epi.4.8.10056

[65] Hou P, Ji M, Xing M. Association of PTEN gene methylation with genetic alterations in the phosphatidylinositol 3-kinase/AKT signaling pathway in thyroid tumors. Cancer. 2008;**113**(9): 2440-2447. DOI: 10.1002/cncr.23869

[66] Deeken-Draisey A, Yang GY, Gao J, Alexiev BA. Anaplastic thyroid carcinoma: An epidemiologic, histologic, immunohistochemical, and molecular single-institution study. Human Pathology. 2018;**82**:140-148. DOI: 10.1016/j.humpath.2018.07.027

[67] Schagdarsurengin U, Gimm O, Dralle H, Hoang-Vu C, Dammann R. CpG island methylation of tumor-related promoters occurs preferentially in undifferentiated carcinoma. Thyroid. 2006;**16**(7):633-642. DOI: 10.1089/thy.2006.16.633

[68] Liu Z, Hou P, Ji M, Guan H, Studeman K, Jensen K, et al. Highly prevalent genetic alterations in receptor tyrosine kinases and phosphatidylinositol 3-kinase/akt and mitogen-activated protein kinase pathways in anaplastic and follicular thyroid cancers. The Journal of Clinical Endocrinology and Metabolism. 2008; **93**(8):3106-3116. DOI: 10.1210/jc.2008-0273

[69] Saghafinia S, Mina M, Riggi N, Hanahan D, Ciriello G. Pan-cancer landscape of aberrant DNA methylation across human tumors. Cell Reports. 2018;**25**(4):1066-1080 e1068. DOI: 10.1016/j.celrep.2018.09.082

[70] dos Reis MB, Barros-Filho MC, Marchi FA, Beltrami CM, Kuasne H, Pinto CAL, et al. Prognostic classifier based on genome-wide DNA methylation profiling in well-differentiated thyroid tumors. Journal of Clinical Endocrinology and Metabolism. 2017;**102**(11):4089-4099. DOI: 10.1210/jc.2017-00881

[71] Kunstman JW, Juhlin CC, Goh G, Brown TC, Stenman A, Healy JM, et al. Characterization of the mutational landscape of anaplastic thyroid cancer via whole-exome sequencing. Human Molecular Genetics. 2015;**24**(8): 2318-2329. DOI: 10.1093/hmg/ddu749

[72] Riesco-Eizaguirre G, Santisteban P. Endocrine Tumours: Advances in the molecular pathogenesis of thyroid cancer: Lessons from the cancer genome. European Journal of Endocrinology. 2016;**175**(5):R203-R217. DOI: 10.1530/EJE-16-0202

[73] Xing M, Tokumaru Y, Wu G, Westra WB, Ladenson PW, Sidransky D. Hypermethylation of the Pendred syndrome gene SLC26A4 is an early event in thyroid tumorigenesis. Cancer Research. 2003;**63**(9):2312-2315. DOI: 10.1097/00130404-200305000-00010

[74] Matsuo K, Tang SH, Zeki K, Gutman RA, Fagin JA. Aberrant deoxyribonucleic acid methylation in human thyroid tumors. Journal of Clinical Endocrinology and Metabolism. 1993;**77**(4):991-995. DOI: 10.1210/jcem.77.4.7691865

[75] Cook M, Yu X-M, Chen H. Notch in the development of thyroid C-cells and the treatment of medullary thyroid cancer. American Journal of Translational Research. 2010;**2**(1): 119-125. DOI: 10.1002/jmv.21734

[76] Mineva I, Gartner W, Hauser P, Kainz A, Löfflfler M, Wolf G, et al. Differential expression of alphaB-crystallin and Hsp27-1 in anaplastic thyroid carcinomas because of tumor-specific alphaB-crystallin gene (CRYAB) silencing. Cell Stress & Chaperones. 2005;**10**(3):171-184. DOI: 10.1379/CSC-107R.1

[77] Vitale G, Caraglia M, Ciccarelli A, Lupoli G, Abbruzzese A, Tagliaferri P, et al. Current approaches and perspectives in the therapy of medullary thyroid carcinoma. Cancer. 2001;**91**(9): 1797-1808. DOI: 10.1002/1097-0142 (20010501)91:9<1797::aid-cncr1199>3.0.co;2-p

[78] Manfredi GI, Dicitore A, Gaudenzi G, Caraglia M, Persani L, Vitale G. PI3K/Akt/mTOR signaling in medullary thyroid cancer: A promising molecular target for cancer therapy. Endocrine. 2015;**48**(2):363-370. DOI: 10.1007/s12020-014-0380-1

[79] Lyra J, Vinagre J, Batista R, Pinto V, Prazeres H, Rodrigues F, et al. mTOR activation in medullary thyroid carcinoma with RAS mutation. European Journal of Endocrinology. 2014;**171**(5):633-640. DOI: 10.1530/EJE-14-0389

[80] Guy GR, Jackson RA, Yusoff P, Chow SY. Sprouty proteins: Modified modulators, matchmakers or missing links? The Journal of Endocrinology. 2009;**203**(2):191-202. DOI: 10.1677/JOE-09-0110

[81] Cerrato A, De Falco V, Santoro M. Molecular genetics of medullary thyroid carcinoma: The quest for novel therapeutic targets. Journal of Molecular Endocrinology. 2009;**43**(4):143-155. DOI: 10.1677/JME-09-0024

[82] Ishida M, Ichihara M, Mii S, Jijiwa M, Asai N, Enomoto A, et al. Sprouty2 regulates growth and differentiation of human neuroblastoma cells through RET tyrosine kinase. Cancer Science. 2007;**98**(6):815-821. DOI: 10.1111/j.1349-7006.2007.00457.x

[83] Macià A, Gallel P, Vaquero M, Gou-Fabregas M, Santacana M, Robledo AMM, et al. Sprouty1 is a candidate tumor-suppressor gene in medullary thyroid carcinoma. Oncogene. 2012;**31**(35):3961-3972. DOI: 10.1038/onc.2011.556

[84] Vitale G, Dicitore A, Pepe D, Gentilini D, Grassi ES, Borghi MO, et al. Synergistic activity of everolimus and 5-aza-2′-deoxycytidine in medullary thyroid carcinoma cell lines. Molecular Oncology. 2017;**11**(8): 1007-1022. DOI: 10.1002/1878-0261.12070

[85] Bellet D, Lavaissiere L, Mock P, Laurent A, Sabourin JC, Bedossa P, et al. Identification of pro-EPIL and EPIL peptides translated from insulin-like 4 (INSL4) mRNA in human placenta. The Journal of Clinical Endocrinology and Metabolism. 1997;**82**(9):3169-3172. DOI: 10.1210/jcem.82.9.4359

[86] Brandt B, Kemming D, Packeisen J, Simon R, Helms M, Feldmann U, et al. Expression of early placenta insulin-like growth factor in breast cancer cells provides an autocrine loop that predominantly enhances invasiveness and motility. Endocrine-Related Cancer. 2005;**12**(4):823-837. DOI: 10.1677/erc.1.00975

[87] Maldonado-Saldivia J, van den Bergen J, Krouskos M, Gilchrist M, Lee C, Li R, et al. Dppa2 and Dppa4 are closely linked SAP motif genes restricted to pluripotent cells and the germ line. Stem Cells. 2007;**25**(1):19-28. DOI: 10.1634/stemcells.2006-0269

[88] John T, Caballero OL, Svobodova SJ, Kong A, Chua R, Browning J, et al. ECSA/DPPA2 is an embryo-cancer antigen that is coexpressed with cancer-testis antigens in non-small cell lung cancer. Clinical Cancer Research. 2008; **14**(11):3291-3298. DOI: 10.1158/1078-0432.CCR-07-1322

Section 4

Experimental Methods in Methylation Researchs

Chapter 7

Atlas of Age- and Tissue-Specific DNA Methylation during Early Development of Barley (*Hordeum vulgare*)

Moumouni Konate, Mike J. Wilkinson, Benjamin T. Mayne, Eileen S. Scott, Bettina Berger and Carlos M. Rodríguez López

Abstract

The barley (*Hordeum vulgare*) genome comprises over 32,000 genes, with differentiated cells expressing only a subset of genes; the remainder being silent. Mechanisms by which tissue-specific genes are regulated are not entirely understood, although DNA methylation is likely to be involved. To shed light on the dynamic of DNA methylation during development and its variation between organs, methylation-sensitive genotyping by sequencing (ms-GBS) was used to generate methylation profiles for roots, leaf-blades and leaf-sheaths from five barley varieties, using seedlings at the three-leaf stage. Robust differentially methylated markers (DMMs) were characterised by pairwise comparisons of roots, leaf-blades and leaf-sheaths of three different ages. While very many DMMs were found between roots and leaf parts, only a few existed between leaf-blades and leaf-sheaths, with differences decreasing with leaf rank. Organ-specific DMMs appeared to target mainly repeat regions, implying that organ differentiation partially relies on the spreading of DNA methylation from repeats to promoters of adjacent genes. Identified DMMs indicate that different organs do possess diagnostic methylation profiles and suggest that DNA methylation is important for both tissue differentiation and organ function and will provide the basis to the understanding of the role of DNA methylation in plant organ differentiation and development.

Keywords: epigenomics, *Hordeum vulgare*, leaf, root, tissue-specific methylation, developmental epigenomics

1. Introduction

DNA methylation is an important characteristic of plant genomes [1, 2], and can occur in all cytosine contexts (CG, CHG and CHH, where H = A, C or T) [3]. The effect of DNA methylation variants on plant development has been demonstrated through methylation alteration tests, which led to plant abnormalities [4, 5]. Furthermore, DNA methylation has been reported to vary from tissue to tissue in

IntechOpen

many species [6–10], and these methylation changes seemed to be essential for normal plant development [11, 12].

Additionally, tissue-specific methylation was proposed to have a strong correlation with the differential expression of some tissue-specific genes. Examples include tissue-specific pigmentation in maize, which is reported to be under epigenetic control [13], and differential gene expression between organs attributed to differentially methylated regions in soybean [14] and sorghum [10]. These studies extended our understanding of the functional importance of tissue-specific DNA methylation, including its role in setting developmental trajectories [9, 13, 15].

A substantial proportion of developmentally expressed genes have alternative promoters (multiple promoters that regulate the same gene) which are under different regulatory programmes [16]. Maunakea et al. [17] proposed that alternative promoters are, at least sometimes, controlled by intragenic DNA methylation. This form of developmental gene regulation is reasoned to be dependent on transposon activity [16] and by implication would mean that silencing of transposons due to DNA methylation may be central to tissue-specific gene expression. Also, tissue-specific gene expression has been associated with methylation changes in promoter regions [2, 18, 19], especially CG islands within promoters [20]. These studies indicate that tissue-specific gene expression does not rely on a single methylation pattern in the genome but, probably, on a combination of variable DNA methylation features.

The magnitude of differential methylation between tissues has been the subject of controversy. It was believed that significant distinctive DNA methylation existed only between specialised tissues such as endosperm, pollen, leaves and roots [9, 10, 21, 22]. Nevertheless, many of these studies also showed that differential DNA methylation between organs, such as roots and leaves, was minor in rice [23], maize [24], sorghum [10] and Arabidopsis [9]. DNA methylation differences between roots and leaves were small in both ^{m}CG and ^{m}CHG contexts [9, 10], with about 1% and 5% divergence, respectively, reported in Arabidopsis [9]. While these studies of differential DNA methylation between tissues generally compared the overall methylation levels [9, 10, 24], these results differ from comparisons made with differentially methylated markers (DMMs) between the same tissues [10], probably due to differences in methylation profiling methods, making it difficult to compare results from different studies. Therefore, it is difficult to know whether differences in the results concerning tissue-specific DNA methylation are due to the plant species or to the approach taken. The study of DNA methylation patterns in plant tissues is important for a better understanding of how these epigenetic markers determine tissue differentiation. Thus, further investigation is warranted to clarify organ specificity of cytosine methylation and the distribution patterns of tissue-specific DNA methylation markers in the plant genome.

To undertake such an investigation, we used barley, a globally important cereal crop, the genome of which has been sequenced recently [25]. The availability of a reference genome made barley a model for the study of cereal crops such as wheat, oats or rye. In this study, we assessed differential DNA methylation between two barley (*Hordeum vulgare*) organs (roots and leaves), using methylation-sensitive genotyping by sequencing (ms-GBS) on five genetically distinct varieties (Barque 73, Flagship, Hindmarsh, Schooner and Yarra). For the sake of simplicity and consistency with the literature, roots and leaves or leaf parts (sheath, blade) may be referred to here as tissues and not organs.

2. Materials and methods

2.1 Plant material and growth conditions

Five spring barley varieties (Barque 73, Flagship, Hindmarsh, Schooner and Yarra), were selected based on their similarity in phenology in order to minimize epigenetic variability between varieties associated with developmental differences. Seeds from all varieties were provided by the Salt Focus Group at the Australian Centre for Plant Functional Genomics (ACPFG, Adelaide, South Australia), and planted at the same time in potting mix comprising 50% UC (University of California at Davis), 35% coco-peat and 15% clay/loam ($v\ v^{-1}$) in 3.3 L pots, 17.5 cm deep, free-draining and placed on saucers. The experiment was conducted from 30th January to 20th February 2015 in a greenhouse at the Waite Campus, University of Adelaide, South Australia (34°58′11″S, 138°38′19″E). The seedlings were grown under natural photoperiod, while temperatures were set at 22°C/15°C (day/night). The experiment consisted of five randomized blocks of five varieties (25 seedlings per block). Pots were watered to weight every 2 days to a gravimetric water content of 16.8% ($w\ w^{-1}$) (0.8 × field capacity) [26] until sampling 21 days after sowing, when seedlings were at three-leaf stage (Zadok stage 13 [27]). Blades and sheaths of leaves 1–3 were sampled separately. Leaves 1 and 2 were fully expanded prior to sampling, whilst leaf 3 had just completed growth. About 50 mg of plant material was cut from the middle section of each leaf blade and each leaf sheath and collected in 2 ml micro tubes. Roots were cut from the seedlings and washed using tap water to remove soil particles, then blotted dry with paper towels before sampling 50 mg of root tissue. All samples were snap frozen in liquid nitrogen, and then stored at −80°C until DNA extraction. In total, 175 tissue samples were collected, including 25 root samples (i.e. 5 plants per each of the five varieties used in the study), 75 leaf blade samples (i.e. from leaves 1, 2 and 3 from each of the 5 plants per variety used in the study) and 75 leaf sheath samples (i.e. from leaves 1, 2 and 3 from each of the 5 plants per variety used in the study).

2.2 DNA isolation

Prior to DNA extraction, frozen plant material was homogenized in a bead beater (2010-Geno/Grinder, SPEX SamplePrep®, USA). DNA isolation was performed from pulverised plant samples using a Qiagen DNeasy kit and following the manufacturer's instructions. DNA samples were quantified using a NanoDrop® 1000 Spectrophotometer (V 3.8.1, ThermoFisher Scientific Inc., Australia) and concentrations were standardized to 10 ng/μl for subsequent library preparation.

2.3 Methylation-sensitive genotyping by sequencing (ms-GBS)

The ms-GBS was performed using a modified version [28] of the original GBS technique [56]. Genomic DNA was digested using the combination of a methylation-insensitive rare cutter, *Eco*RI (GAATTC), and a frequent and methylation-sensitive cutter, *Msp*I (CCGG). Each sample of DNA was digested in a reaction volume of 20 μl containing 2 μl of NEB Smartcut buffer, 8 U of HF-*Eco*RI (High-Fidelity) and 8 U of *Msp*I (New England BioLabs, Australia). The reaction was performed in a BioRad 100 thermocycler at 37°C for 2 h, followed by enzyme inactivation at 65°C for 10 min.

Then the ligation of adapters to individual samples was achieved in the same plates by adding 0.1 pmol of the respective barcoded adapters with an *Msp*I cut site overhang, 15 pmol of the common Y adapter with an *Eco*RI cut site overhang, 200 U of T4 Ligase and T4 Ligase buffer (New England BioLabs, Australia) in a total volume of 40 μl. Ligation was carried out at 24°C for 2 h followed by an enzyme inactivation step at 65°C for 10 min.

DNA samples were allocated to plates, 81 samples each, including the negative control, water. Prior to pooling plate samples into a single 81-plex library, the ligation products were individually cleaned up to remove excess adapters using an Agencourt AMPure XP purification system (Beckman Coulter, Australia) at a ratio of 0.85 (AMPure magnetic beads/ligation product), following the manufacturer's instructions. Individual GBS libraries were produced by pooling 25 ng of DNA from each sample. Each constructed library was then amplified in eight separate PCR (25 μl each) containing 10 μl of library DNA, 5 μl of 5× Q5 high fidelity buffer, 0.25 μl polymerase Q5 high fidelity, 1 μl each of Forward and Reverse common primers at 10 μM, 0.5 μl of 10 μM dNTP and 7.25 μl of sterile pure water. PCR amplification was performed in a BioRad T100 thermocycler, consisting of DNA denaturation at 98°C (30 s) and 10 cycles of 98°C (30 s), 62°C (20 s) and 72°C (30 s), followed by 72°C for 5 min. PCR products were next pooled to reconstitute libraries. DNA fragments between 200 and 350 bp in size were captured using AMPure XP magnetic beads following the manufacturer's instructions. Bead-captured fragments were eluted in 35 μl of water, of which 30 μl were collected in a new labelled microtube. Libraries were next paired-end sequenced in an Illumina HiSeq 2500 (Illumina Inc., USA) at the Australian Genome Research Facility (AGRF, Melbourne Node, Australia). Sequencing results were deposited in the European Nucleotide Archive (ENA) (Study Accession Number: PRJEB27251).

2.4 Analysis of global differences in DNA methylation between samples

Differences in ms-GBS profiles between samples were explored by performing principal component-linear discriminant analysis (PC-LDA) (a supervised clustering approach for high dimensional data), using different levels of hierarchy between samples as the putative drivers in DNA methylation differences (i.e. grouping samples by organ (root vs. leaf), tissue (root vs. blade vs. sheath, and tissue) and age (root vs. leaf 1 vs. leaf 2 vs. leaf 3 vs. sheath 1 vs. sheath 2 vs. sheath 3)). PC-LDA was implemented using the R package FIEmspro 1.1-0 [29] on the standardized coverage, the count per million reads (CPM) of the 913,697 ms-GBS markers generated. PC-LDA results were visualized by a scatter plot of the first two discriminant factors (DFs), and a 3D plot using the first three DFs. Finally we used an unsupervised hierarchical cluster analysis to generate a dissimilarity tree based on Mahalanobis distance [30] generated also based on the standardized coverage (CPM) of the 913,697 ms-GBS markers.

2.5 Detection of DMMs in barley

Differentially methylated DNA was assessed in ^{m}CCGG motifs (recognised by *Msp*I), between barley leaf parts (blade and sheath) and roots. To do so, samples were grouped according to organ type (root, blade and sheath) regardless of the genotype of origin, making 25 samples per organ. This approach aimed to minimise genotype-dependent methylation markers. DMMs were identified using the package, *msgbsR*, developed by Mayne et al. [31]. DMMs were selected based on FDR adjusted P-values with a threshold of 0.05 [32, 33]. The significance of the marker also fulfilled the condition that the read counts reached at least 1 CPM and was present in at least 20 samples per organ type (maximum sample per group = 25). The *log*FC (logarithm 2 of fold-change) was computed to estimate the intensity and directionality of differential DNA methylation between tissues. Determining the directionality of DNA methylation uses the fold change as an inverse proxy for change in the methylation level. That is, higher methylation levels on a specific locus will reduce the number of *Msp*I restriction products and therefore reduce the number of sequences generated for that locus [34].

2.6 Distribution of DMMs around genomic features

To test whether there was a relationship between tissue-specific DMMs and particular genomic features (e.g., genes and repeat regions as defined in Ensembl database (http://plants.ensembl.org/biomart/martview/)), DMM distribution was assessed in the barley genome. Therefore, DMMs stable between tissues were mapped to the barley reference genome. Then, the number of DMMs within genomic features (repeats, genes, exons, UTRs and tRNA genes) and per 1 kb bins within 5 kb flanking regions [24, 28] was tallied using the shell module, *bedtools/2.22.0* [35].

3. Results

3.1 Methylation-sensitive genotyping by sequencing

The sequencing of the 170 samples of barley tissue which met DNA quality requirements yielded over 900 million raw reads, with more than 91% bases above Q30 (99.9% accuracy of base call [36]) across all samples (**Table 1**). Of these reads, 99.27% contained the barcode and *Eco*RI/*Msp*I adapters ligated during library construction. Further filtering was performed to retain reads that strictly aligned with the barley reference genome. In this way, we obtained nearly 450 million reads (50.10%), with a mean of 2,637,916 high quality reads per sample. These high-quality reads accounted for 913,697 sequence tags, representing 32.30% of the 2,828,642 CCGG sites in the barley genome. Of these sequence tags, 748,594 (80.62%) showed some form of polymorphism for methylation between samples.

3.2 Estimation of tissue- and tissue rank-dependent epigenetic differentiation

The PC-LDA plots revealed clear evidence of structuring of methylation between samples (**Figure 1a**). A 3D plot using the first three discriminant factors

Sequencing parameters	Yield
Raw reads	901,617,058
Reads that matched barcodes	895,013,295
Reads aligned to barley reference genome	448,445,748
Samples	170
Average reads per sample	2,637,916
Total unique tags	913,697
Polymorphic tags	748,594

Table 1.
Data yields from ms-GBS, generated using the Illumina HiSeq 2500 platform.

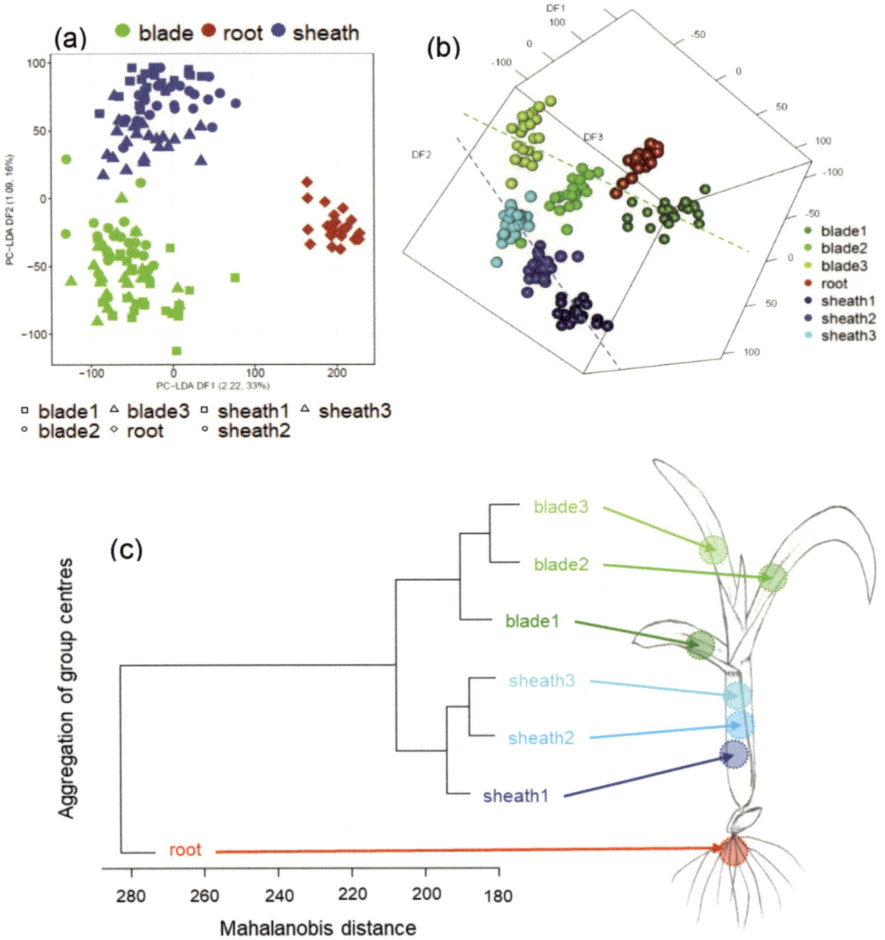

Figure 1.
Analysis of the differentiation of DNA methylation profiles of barley roots, leaf sheaths and leaf blades. (a) Scatter plot of the first two discriminant factors of the principal component-linear discriminant analysis (PC-LDA) (DF1 and DF2) using 913,697 ms-GBS markers generated from genomic DNA of roots, leaf sheaths and leaf blades, collected from 25 barley plants at the three-leaf stage (21 days after sowing), comprising five varieties (Barque 73, Flagship, Hindmarsh, Schooner and Yarra). (b) Three-dimensional plot of the first three discriminant factors of the PC-LDA of the same ms-GBS data. (c) Hierarchical cluster of the distances between sample group centres, based on Mahalanobis distance. Blade 1-3 and sheath 1-3 indicate the rank of the organ type, first, second and third leaf of seedlings, respectively.

(DF1, DF2 and DF3) revealed that blades and sheaths were further grouped according to the rank of the leaf from which they were harvested. The distance between blades and sheaths seems to shrink with leaf rank (**Figure 1b**). This leaf rank-dependent grouping was also supported by hierarchical cluster analysis (HCA) of the distances between sample group centres (**Figure 1c**), based on the Mahalanobis distance [29, 30], and sample clusters matched the leaf developmental age (**Figure 1c**). Leaf rank-dependent DNA methylation differences were further assessed between tissues by comparing the methylation profiles of blades and sheaths for each rank of leaf appearance. No DMMs were observed between the three leaf blades, whereas sheaths 1 and 3 presented 18 DMMs (**Table 2**).

3.3 Differentially methylated DNA markers between roots and leaves

DMMs between barley roots and leaves were obtained through comparison of the read count per million of tissue types, independently of genotypes.

	Blade 1	Blade 2	Blade 3	Sheath 1	Sheath 2	Sheath 3
Blade 1	—					
Blade 2	0	—				
Blade 3	0	0	—			
Sheath 1	32	37	73	—		
Sheath 2	29	36	40	0	—	
Sheath 3	0	1	1	18	0	—

Differentially methylated markers (FDR <0.05) were obtained from 913,697 ms-GBS tags generated from genomic DNA of barley roots, leaf sheaths and leaf blades, collected from 25 plants at three-leaf stage (21 days after sowing) of five barley varieties (Barque 73, Flagship, Hindmarsh, Schooner and Yarra). Blade 1–3 and sheath 1–3 indicate the rank of the leaf; first, second and third, respectively, on seedlings.

Table 2.
Number of differentially methylated markers in barley tissues of different ages.

This comparison revealed substantial DMMs between both roots vs. blades and roots vs. sheaths (**Figure 2a**), and there were more DMMs between roots and blades (6510 DMMs **Figure 2b**) than between roots and sheaths (4116 DMMs **Figure 2c**). Of these markers, 3266 DMMs were present in both blades and sheaths when compared to roots, and their methylation changed consistently in the same direction in each comparison (**Figure 3a**). The number of DMMs between roots and leaf blades increased with leaf-rank, whereas DMMs between roots and leaf sheaths did not show any relationship with rank (**Figure 2a**). Tissue-specific DMMs were predominantly hypomethylated (95–98%) in leaf parts (sheath or blade) compared to roots (**Figure 2a**). This result was in line with the median of the fold-changes of DMMs, which indicated an overall DNA hypomethylation in leaves (**Figure 4a** and **b**). From here on, DMMs consistently present in roots vs. sheaths and roots vs. blades will be designated as stable markers between roots and leaves.

3.4 Differentially methylated DNA markers between the leaf blade and sheath

There was only a small number of DMMs between leaf blades and sheaths (0–73 DMMs, **Table 2** and **Figure 2d**). These DMMs were basically between leaf blades and sheaths 1 and 2; and there was none between blade 1 and sheath 3. There was only 1 DMM between sheath 3 and blades 2 and 3 (**Table 2** and **Figure 2d**). Pairwise comparisons between blades 1–2 and sheaths 1–2 revealed 20 common DMMs, which were all hypermethylated in sheaths compared to blades (**Figures 2e** and **4b**). Half of the 20 common DMMs between blades and sheaths were located on chromosome 5H. Furthermore, there were no DMMs in pairwise comparisons among blades 1–3 and among sheaths 1–3, except between sheath 1 and sheath 3 which had 18 DMMs (**Table 2**). However, comparing blades and sheaths of the same leaf rank showed 32 DMMs between blade 1 and sheath 1, 36 DMMs between blade 2 and sheath 2 and 1 DMM between blade 3 and sheath 3.

3.5 Distribution of tissue-specific DMMs around genes

Relatively few of the tissue-specific DMMs were located around gene exons. Indeed, of the 3266 stable DMMs between root and leaf samples, only 60 (1.8%) were located within 5 kb of a gene, including 21 overlaps with genes and 39 DMMs that were spread within 5 kb upstream and downstream of genes (**Figure 5a**). Apart from the absence of DMMs within 1 kb upstream of transcription start sites, there was no obvious tissue-specific DMM distribution pattern around the genes

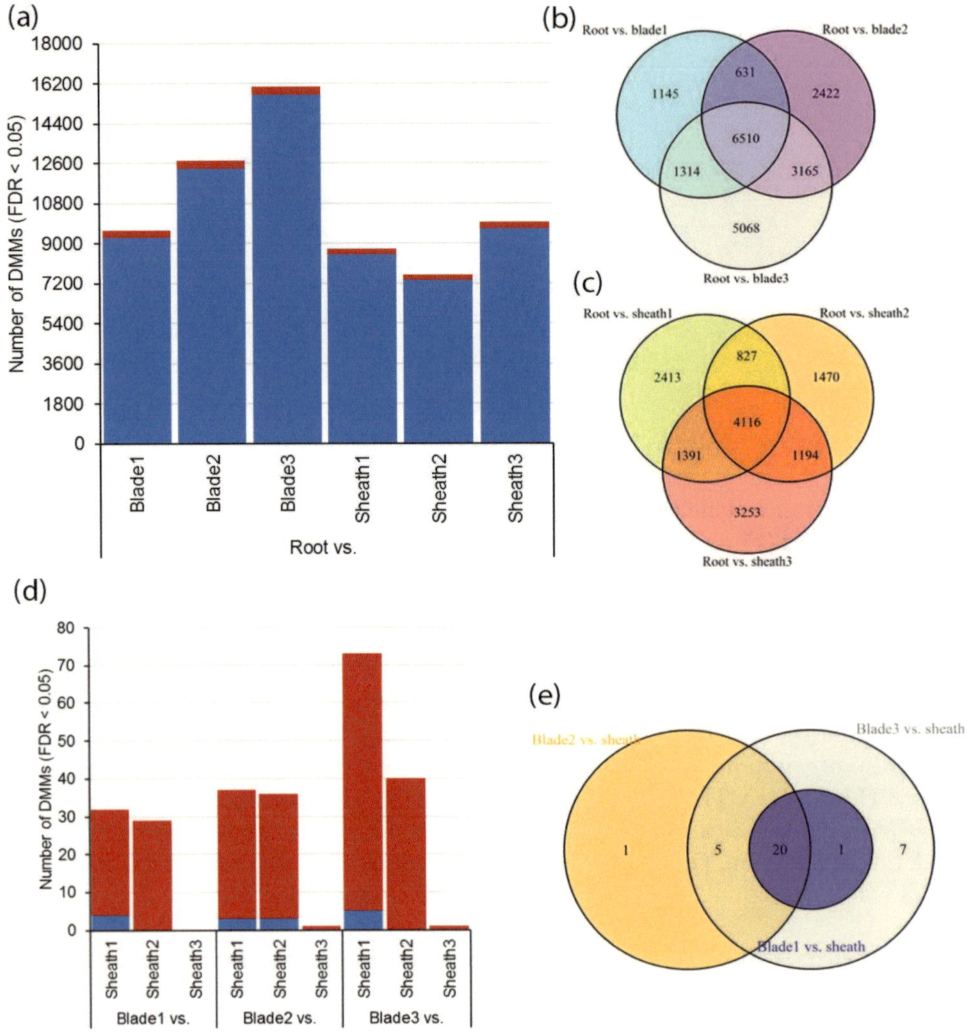

Figure 2.
Analysis of the number of DMMs among three barley tissues. (a) Number of DMMs between roots and leaf blades (root vs. blade) and roots and sheaths (roots vs. sheaths). Histogram colour indicates whether the DMMs are hypomethylated (blue) or hypermethylated (red) in leaf parts compared to roots. (b and c) Venn diagram showing the number of DMMs stable between root and blade tissues (b) and between root and sheath tissues (c). (d) Number of DMMs from pairwise comparison between leaf blades 1–3 and sheaths 1–3. Histogram colour indicates whether the DMMs are hypomethylated (blue) or hypermethylated (red) in sheaths compared with blades. (e) Venn diagram showing the number of DMMs common in pairwise comparisons between leaf blades 1–3 and sheaths 1–2. Tissue samples were collected from seedlings at the three-leaf stage of five barley varieties grown in five replicates for 21 days after sowing. Blade 1–3 and sheath 1–3 indicate the rank of the organ type; first, second and third, respectively, on seedlings. DMMs were selected based on the significance of the false discovery rate, FDR, <0.05. DMMs present in both sheaths and blades when compared with roots, have been designated as markers between roots and leaves.

(**Figure 5a**). The same assessment process showed that, as with common DMMs, only a small proportion of blade-specific DMMs (44 of 3246, 1.3%) was positioned close to a gene (**Figure 5b**). Of these, 15 DMMs overlapped with a gene transcript, whereas the remaining 29 DMMs were distributed within 5 kb of the gene without any clear pattern (**Figure 5b**), except that the number of DMMs located between 2 and 3 kb bins was higher both upstream and downstream, than any other 1 kb bin within the 5 kb flanking regions (**Figure 5b**). There were fewer sheath-specific methylation markers within 5 kb from genes than blade-specific markers (13 of

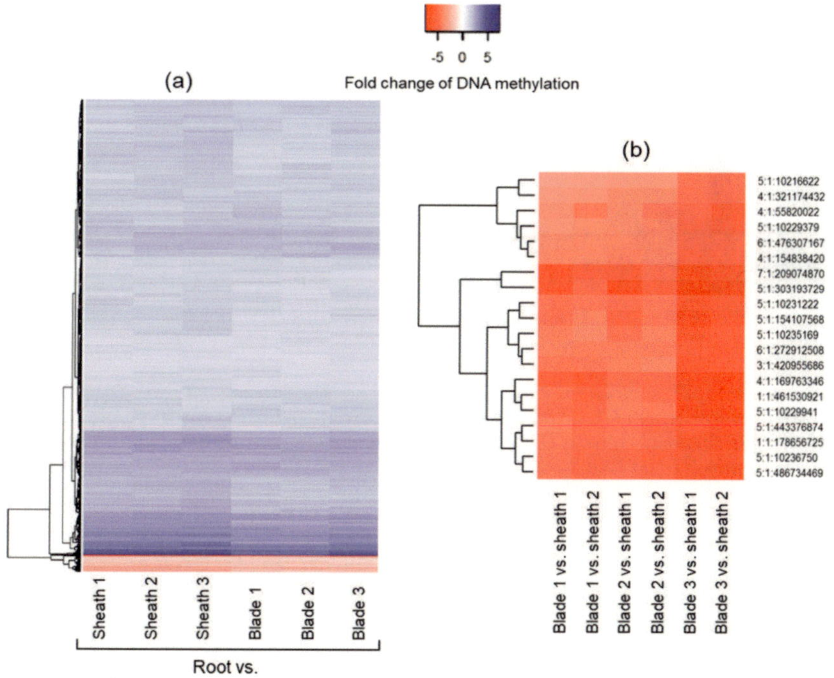

Figure 3.
Hierarchical clustering analysis of the DMMs. (a) The 3266 common DMMs between roots and all leaf parts (sheath 1–3, blade 1–3). The colours in the heat map indicate whether the DMM is hypomethylated (blue) or hypermethylated (red) in leaf parts compared to roots. (b) Hierarchical clustering of the 20 stable DMMs between blades and sheaths. In this heat map the red colour shows hypermethylation of DMMs in sheaths compared with blades. Blade and sheath samples were collected from seedlings at three-leaf stage of five barley varieties grown in five replicates for 21 days after sowing. Blade 1–3 and sheath 1–3 indicate the rank of the leaf on seedlings, first, second and third, respectively. The first number of the marker label on the y axis indicates the chromosome number on which the marker is located.

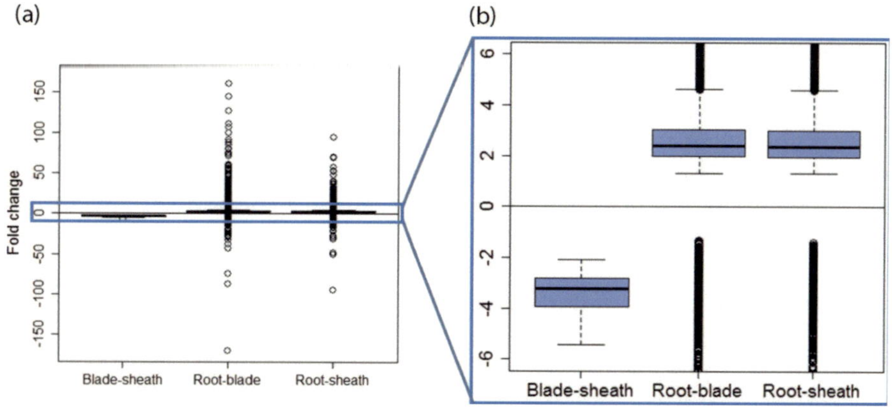

Figure 4.
Directionality of the methylation in tissue-specific DNA methylation markers. (a) Boxplots showing the spread of the fold-change of locus read counts between blades and sheaths, roots and blades, and roots and sheaths. (b) Detail of boxplots, highlighting the median of methylation fold-change of all loci in each comparison. The fold-change of DNA methylation was estimated by computing $2^{(log^{2FC})}$*, with* log2FC *= logarithm 2 of fold-change in read counts for each sequenced locus between pairwise comparisons of tissues collected from three-leaf stage barley seedlings. Leaf blades were the reference state for blade-sheath comparison, whereas roots were the reference for root-blade and root-sheath comparisons. Negative and positive values on the y axis indicate respectively, hypermethylation and hypomethylation of the tissue that is compared to the reference. Locus coverage was estimated for each tissue by using 25 replicates for roots and 75 for blades and sheaths (5 plants from each of the 5 varieties included in the study (DNA was extracted from 1 single root and from 3 independent leaves per plant)).*

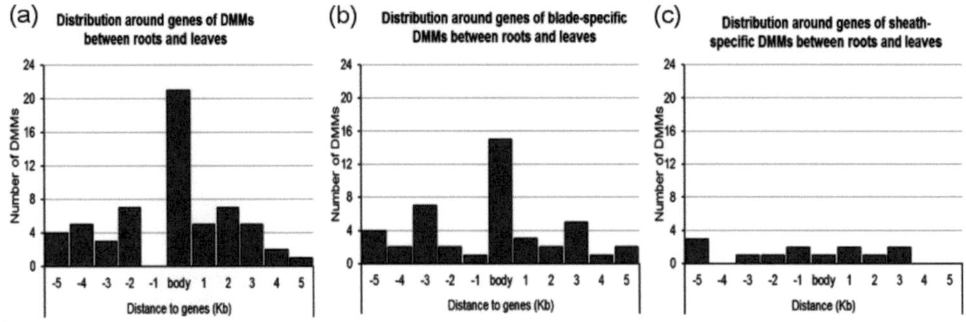

Figure 5.
Distribution of tissue-specific differentially methylated markers (DMMs) around genes. (a) DMMs between roots and leaves, present in both blades and sheaths as in **Figure 2b** *and* **c***; (b) blade-specific DMMs between roots and leaves and (c) sheath-specific DMMs between roots and leaves. The y axis indicates the distance to genes in kilo base pairs (kb) on both flanking regions. Negative and positive values indicate upstream and downstream of genes, respectively. DMMs overlapping with genes are considered as changes in gene-body methylation (body). The x axis shows the number of DMMs per 1 kb window.*

2391 DMMs, 0.5%) (**Figure 5c**). The majority of these (10 out of 13 DMMs) were sited within 3 kb of a gene, and no DMMs were present 3–5 kb from transcription margins (**Figure 5c**). Of 37 gene-body DMMs detected across all comparisons (**Figure 5a–c**), 27 overlapped with an exon and the remaining 10 markers were in intronic regions, 70–604 bp upstream of exons, except 1 DMM, which was 62 bp downstream an exon (Appendix A).

3.6 Distribution of tissue-specific DMMs near repeat regions

Many more tissue-specific DMMs were detected near repeats than near genes. The DMMs around repeat regions (as defined in the Ensembl database (http://plants.ensembl.org/biomart/martview/)) were concentrated either within the repeats or within 1 kb of their margins (**Figure 6a**). A similar distribution pattern was obtained with both blade-specific and sheath-specific DMMs when contrasted with roots, with more DMMs overlapping with the repeats themselves than in the 1 kb stretches flanking their margins (**Figure 6b** and **c**). The few markers that were differentially methylated between blades and sheaths (20 DMMs in total) were all located within 1 kb of a repeat (**Figure 6d**). Therefore, stable tissue-specific DMMs appeared to occur preferentially within repeats and 1 kb flanking regions, with higher frequency within 1 kb downstream than within 1 kb upstream, regardless of tissue types (**Figure 6a–d**).

3.7 Distribution of genes around differentially methylated (DM) repeats

To investigate a possible interaction between differentially methylated (DM) repeats and genes, the distance of genes from DM repeats between root and leaf samples was evaluated. In this way, we found 105 genes near repeats (up to 5 kb either side), of which 37 overlapped with a repeat and the remaining genes were scattered up- and downstream from the repeat (**Figure 7**). The number of DM repeats surrounded by genes thus represented only a tiny proportion of the total repeats that were differentially methylated between roots and leaves (105 out of 3266 DM repeats, 3.21%). About half of genes near DM repeats (52 of 105 genes) were also differentially methylated, whereas the remainder (53 genes) were not.

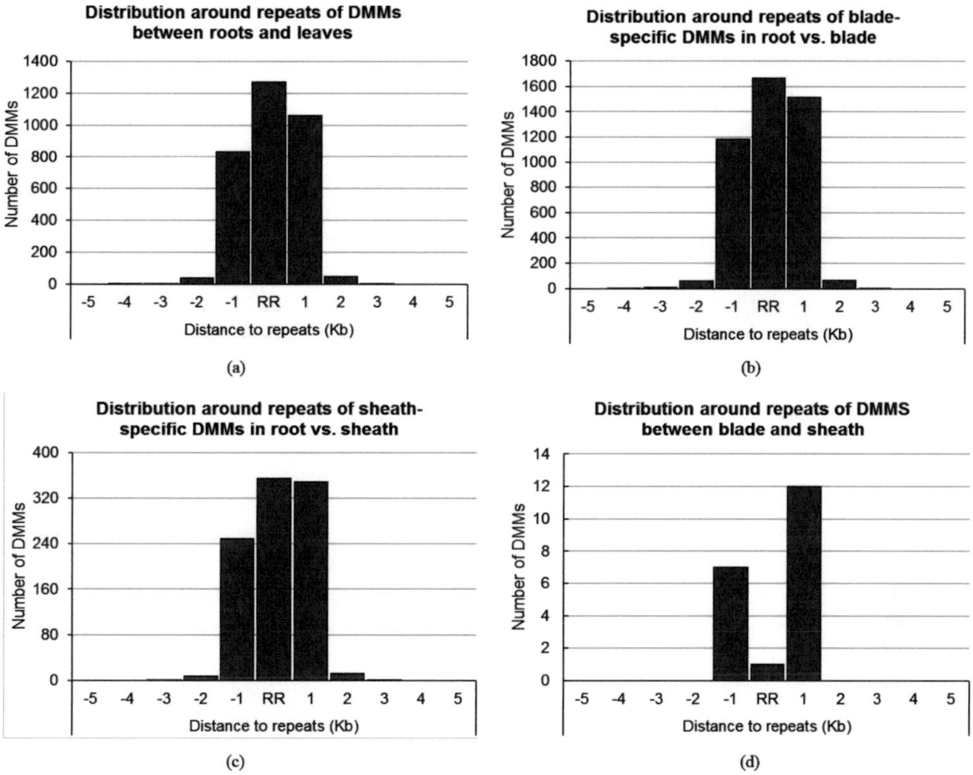

Figure 6.
Distribution of tissue-specific differentially methylated markers (DMMs) around repeats. (a) DMMs between roots and leaves, present in both blades and sheaths as in **Figure 2b** *and* **c**; *(b) blade-specific DMMs between roots and leaves; (c) sheath-specific DMMs between roots and leaves; (d) DMMs between blades and sheaths. The x axis indicates the distance to repeats in kilo base pairs (kb) on both flanking regions. Negative and positive values indicate upstream and downstream repeat regions, respectively. RR: repeat regions. The y axis shows the number of DMMs per 1 kb window.*

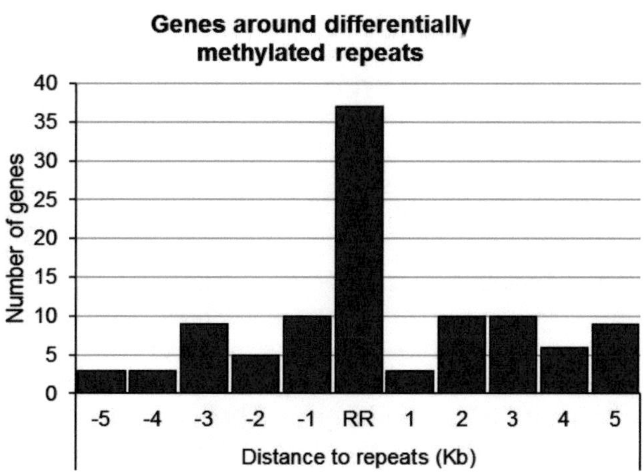

Figure 7.
Distribution of genes around differentially methylated repeat regions. The x axis indicates the distance to repeats in kilo base pairs (kb) on both flanking regions. Negative and positive values indicate upstream and downstream repeat regions, respectively. RR, repeat regions. The y axis shows the number of genes per 1 kb window.

4. Discussion

4.1 Extensive epigenetic differentiation between roots and leaves

In this study, we detected large numbers of DDMs between roots and leaves that were conserved across a diverse array of barley genotypes, and so were deemed far more likely to be organ-specific than genotype-dependent. Of these, hypomethylation of the ^{m}CCGG motif predominated in leaves (**Figures 2b** and **c**, **3b** and **4a**). More surprisingly, we also detected similarly conserved DMMs between leaf-blades and leaf-sheaths (**Figures 2e** and **4b**). The number of conserved DMMs between blades and sheaths (20 DMMs), all hypermethylated in sheaths, was relatively consistent with the closeness of these structures in position and function. These findings are broadly congruent with previous studies, which reported differential DNA methylation between variable tissues (e.g. endosperm, pollen, leaves, and roots) in diverse plant species [7–10], but additionally hint that the developmental closeness of structures being compared may also be reflected in the distinctiveness of their methylation profiles. However, controversy over the extent and validity of organ-specific DMMs [9, 10, 21–23] could cast doubt over their utility for organ diagnosis or as a tool to gain greater insight into the genes responsible for organ development/identity. Here, we sought to mitigate against the possibility of type I errors in DMM assignment through the unprecedented use of five diverse varieties and five biological replicates of each variety in the identification of these marks. In contrast to our findings, previous workers have reported little difference in the methylation levels of both ^{m}CG and ^{m}CHG motifs between roots and leaves in Arabidopsis [9] and sorghum [10]. Further, no significant difference was detected at all for ^{m}CG and ^{m}CHG methylation levels between tissues in cotton [37]. These divergences may simply reflect genuine biological differences between taxonomic groups. However, it is also important to recognise that such differences may also arise from the approach used to identify organ-specific DMMs. Variability in the techniques used to assess plant methylation profiles may introduce different forms of bias and preclude or complicate comparison among studies. DMM detection can be influenced by factors such as (1) the genome coverage of the methylation profiling method (low coverage methods such as MSAP are likely to miss many markers) [7], and (2) the data analysis approach used, which can compare either global methylation levels (e.g. percent methylation) [9] or methylated loci (e.g. DMMs) [28]. We contend that relying solely on global methylation levels can be misleading in comparing tissue profiles, because similar methylation levels may show completely different patterns and so vital information content is lost.

The current study revealed that tissue-specific DNA methylation occurred abundantly in the ^{m}CHG context (in particular ^{m}CCGGs) (**Figure** 2a and **c**). This concurs with reports of the CHG context similarly dominating differential DNA methylation between organs in *Brachypodium distachyon* [8] and sorghum [10]. Although tissue-specific methylation also occurs in other cytosine contexts [10], our results and other studies [10, 22] suggest that ^{m}CCGG is a primary motif of epigenetic distinctiveness of plant organs. However, since *Msp*I activity is affected by the presence of cytosine hydroxymethylation on its recognition sequence [38], some of the markers identified here as being cytosine methylation induced, could be due to the presence of (de) hydroxymethylation events instead. Additionally, while tissue-specific DMMs were mostly hypomethylated in leaves compared to roots in the present study (**Figure 3b**), in Arabidopsis, Widman et al. [9] found that hypermethylation prevailed in leaves compared with roots. This apparent contradiction in the directionality of methylation in DMMs between roots and leaves may be a reflection a difference in the polarity of early divisions in the monocotyledonous barley and the dicotyledonous Arabidopsis embryos or else the methylation profiling method implemented.

4.2 DNA methylation flux is tissue-specific during barley seedlings development

In addition to tissue-specificity of methylation profiles, one notable finding in the current study was that leaf cohorts exhibited a strong tendency to co-cluster. This suggests that the nature of methylation divergence between organs is not absolutely fixed and instead appears to change with developmental progression. This observation accords with previous reports that genome-wide methylation patterns are not static during plant development [39]. Additionally, a considerable portion of DMMs between roots and leaves was also specific to the leaf rank, due to the steady decrease in the number of DMMs between roots and leaf blades with the rank of the latter (**Figure 2a–c**). In this case, therefore, the slow but progressive accumulation of additional methylation marks in the leaves increases their divergence from root profiles and enables the separation of leaf cohorts. However, the small number of DMMs distinguishing between leaf blades and leaf sheaths ran counter to this trend such that there were no DMMs capable of discrimination between these leaf parts among the oldest cohort studied (leaf 1) (**Figure 2d** and **Table 2**). It seems intuitively improbable that older cohorts of leaves would simply lose differentiation between structurally distinct parts, especially if these marks had a functional role in defining function. Perhaps the most plausible biological explanation for the apparent erosion of divergence lies in the different chronological ages of the leaf cohorts that were sampled. Put simply, the third leaves were the least mature of the three cohorts collected and so it is entirely possible that the blade-sheath differential marks had yet to appear in these samples. Thus, it is important to consider the developmental and ageing progression chronology when assigning DMMs and that some organ- or structure-specific marks may only become organ-specific late in their development. Such late-emerging developmental DMMs should mean that the cumulative number of tissue-specific markers increases and so the organs or structures become more distinct, through leaf growth stages [40], each of which may carry a specific epigenetic profile. Certainly, others have noted that methylation profiles vary progressively as the organ develops [3, 41, 42] before reaching, at maturity, a "default" methylome, which may be conserved across varieties [24]. These results suggest that, once leaves are differentiated and mature, they do not show significant differences in DNA methylation profiles, regardless of their rank of appearance. Additionally, the location of half of the 20 common DMMs between blades and sheaths on chromosome 5H implies that this chromosome carries loci important for blade and sheath identities.

4.3 Tissue-specific DNA methylation preferably occurs in repeat regions of the barley genome

Organ-specific DMMs identified here were primarily associated with repeat regions. No significant difference was observed between the frequency of CCGG sites in and around genes and repeats. However, 84% of the barley genome is comprised of mobile elements or other repeat structures [25, 43], indicating that the fact that the majority of detected DMMs are located within or in the proximity of a repeat is due to the intrinsic repetitive nature of the studied genome. Nevertheless, the fact that 27 DMMs overlapped with exons and 10 were located in introns (Appendix A) contradicts previous claims that CHG methylation marks are exclusively restricted to repeat regions and intergenic regions [20, 21, 44, 45]. The possible regulatory significance of such gene body CHG methylation marks requires further investigation [46]. However, it is already well-established that tissue-specific DMMs can influence gene expression by enhancing gene transcription [9] and alternative splicing [47] or through repression due to immediate proximity to transcription start site [48].

The predominance of DMMs around and within repeats leads us to speculate that they could play an important role in defining organ identity in barley, and accords with previous findings in *Brachypodium distachyon* [8]. This flux of DNA methylation patterns in repeats [8, 42, 49] has been proposed to regulate [44] developmental shifts during plant growth and development [11, 39]. Nevertheless, the association between DMMs in/around repeat regions and organ identity described here does not establish a causal link between the two. However, there are grounds for reasoning that this may be the case and that the possibility warrants further study. First, repeat regions were previously proposed to be involved in alternative promoters, a substantial proportion of which (>40%) was reported to shape tissue differentiation [16]. Therefore, tissue-specific DMMs in repeats may contribute to alternative promoters, and thus determine organ identity. Second, differential gene expression between roots and leaves [25, 50] implies a firm regulatory system, including epigenetic mechanisms to guarantee tissue-specific cell development. Tissue-specific DMMs in repeats show that repeats are not the so-called "selfish parasites" of the genome [51], but can directly or indirectly affect tissue-specific gene expression [42, 52, 53]. Finally, it has been suggested that transposons coordinate splice variants, a genomic event that occurs in more than 60% of plant genes [54, 55], thus generating multiple mRNA transcripts from a single gene [56, 57]. Many splice variants are tissue-specific [58], suggesting that it is entirely possible that tissue-specific DMMs in repeats affect alternative splicing and subsequent gene expression. Also, some DM genes might potentially be regulated simultaneously by their own methylation and that of repeats [53, 59], due to proximity with DM repeats.

5. Conclusions

This study provides a comprehensive set of robust tissue specific epimarkers which were conserved in all barley genotypes tested and can therefore be considered genotype independent. Such markers have potential to be converted into locus-specific methylation sensitive cleaved amplified polymorphic sequence markers (ms-CAPS) to be used as diagnostic of sample origin. Moreover, these markers provide a basis for the understanding of the role of DNA methylation in plant organ differentiation and development. Our data illustrates that during tissue development, DNA methylation evolves to reach a default profile once the tissue is completely differentiated at maturity. It is possible that the plant organ formation and maturation is under at least partial control of DNA methylation changes. In addition, repeats could play an important role in tissue definition. The existence of tissue-specific mCCGG sites suggests that this context carries important factors of tissue differentiation. Expression analysis of tissue samples would conclusively demonstrate the role of tissue-specific DMMs in gene regulation. These markers will provide a basis for future studies of the role of DNA methylation in plant organ differentiation and development.

Acknowledgements

The authors are grateful to the Australian Agency for International Development (AusAID) for supporting MK with an Australian Awards Scholarship. MJW was paid by the BBSRC (BBS/E/W/0012843C). This work is supported by the National

Institute of Food and Agriculture, U.S. Department of Agriculture, Hatch Program number 2352987000, and award number 2019-67013-29168.

Conflict of interest

"The authors declare no conflict of interest."

Author contributions

MK conceived and performed the experiments, analysed the data and wrote the manuscript; BJM performed ms-GBS data alignments; MJW, ESS, BB, CMRL conceived the experiments and supervised the work. All authors read and commented on the manuscript.

Appendix

Chrom.	Exons				DMMs			Tissue
	Start	End	ID	Rank	Start	End	bp to exon	
3H	256588863	256589313	exon:MLOC_37071.2:3	3	256588258	256588258	−604	leaf
1H	173809114	173809167	exon:MLOC_44613.1:2	2	173808619	173808619	−494	leaf
2H	**427507334**	**427507612**	**exon:MLOC_61110.4:1**	**1**	**427506881**	**427506881**	**−452**	**blade**
7H	584462328	584462663	exon:MLOC_6930.1:4	4	584461984	584461984	−343	blade
3H	48188588	48188710	exon:MLOC_36518.3:9	9	48188256	48188256	−331	leaf
4H	531043445	531043540	exon:MLOC_66787.2:5	5	531043255	531043255	−189	leaf
3H	282775878	282775978	exon:MLOC_57866.1:2	2	282775689	282775689	−188	leaf
2H	507101612	507102232	exon:MLOC_57766.1:6	6	507101429	507101429	−182	blade
3H	451801679	451801792	exon:MLOC_4568.8:12	12	451801608	451801608	−70	blade
1H	295869691	295869957	exon:MLOC_57040.1:1	1	295869907	295869907	0	blade
1H	372664328	372665243	exon:MLOC_11591.1:1	1	372665217	372665217	0	leaf
1H	398203764	398206694	exon:MLOC_52730.3:1	1	398204886	398204886	0	leaf
2H	436039625	436040167	exon:MLOC_16240.2:1	1	436040156	436040156	0	leaf
2H	550574223	550574658	exon:MLOC_7365.2:1	1	550574622	550574622	0	leaf
3H	141116151	141117572	exon:MLOC_70576.2:1	1	141116946	141116946	0	blade
4H	428185287	428190462	exon:MLOC_52907.1:1	1	428185685	428185685	0	leaf
5H	449547966	449548309	exon:MLOC_66740.1:1	1	449548006	449548006	0	blade
6H	5471445	5474755	exon:MLOC_54256.1:1	1	5473235	5473235	0	leaf
6H	247447067	247450327	exon:MLOC_7517.2:1	1	247448194	247448194	0	blade
7H	96048516	96048816	exon:MLOC_36488.1:1	1	96048734	96048734	0	leaf
7H	440064807	440067513	exon:MLOC_72767.1:1	1	440065330	440065330	0	leaf
7H	544501261	544504310	exon:MLOC_39738.1:1	1	544501865	544501865	0	sheath
6H	69839676	69839776	exon:MLOC_11882.4:2	2	69839743	69839743	0	leaf
7H	331094393	331097017	exon:MLOC_54330.1:2	2	331096165	331096165	0	blade
1H	61790876	61791279	exon:MLOC_66388.8:3	3	61791253	61791253	0	leaf
3H	421991486	421991892	exon:MLOC_18521.3:3	3	421991580	421991580	0	leaf

Chrom.	Exons				DMMs			Tissue
	Start	End	ID	Rank	Start	End	bp to exon	
7H	96049105	96050237	exon:MLOC_36488.1:3	3	96049134	96049134	0	leaf
3H	516390233	516390451	exon:MLOC_37766.1:4	4	516390244	516390244	0	blade
4H	434415593	434415838	exon:MLOC_58529.1:4	4	434415773	434415773	0	blade
2H	578608506	578608551	exon:MLOC_54514.1:5	5	578608549	578608549	0	blade
5H	484203288	484203413	exon:MLOC_73139.2:5	5	484203386	484203386	0	blade
2H	2183704	2183865	exon:MLOC_57446.2:9	9	2183753	2183753	0	leaf
7H	41386814	41387497	exon:MLOC_57450.2:9	9	41387134	41387134	0	leaf
4H	434420196	434420586	exon:MLOC_58529.6:13	13	434420355	434420355	0	blade
3H	541205210	541205401	exon:MLOC_37244.3:16	16	541205351	541205351	0	leaf
7H	570620131	570620572	exon:MLOC_14604.2:16	16	570620258	570620258	0	blade
7H	583930566	583930636	exon:MLOC_62970.1:2	2	583930697	583930697	62	leaf

DMMs: differentially methylated markers; Chrom: chromosome; bp: base pair

Table A1.
List of differentially methylated exons. Bolded value is the only first exon methylated upstream 452 bp from a transcription start.

Author details

Moumouni Konate[1], Mike J. Wilkinson[2*], Benjamin T. Mayne[3], Eileen S. Scott[4], Bettina Berger[5] and Carlos M. Rodríguez López[6*]

1 Institut de l'Environnement et de Recherche Agricole (INERA), Station de Farako-Ba, Burkina Faso

2 Pwllpeiran Upland Research Centre, Institute of Biological, Environmental and Rural Sciences, Aberystwyth, Ceredigion, UK

3 Robinson Research Institute, School of Medicine, The University of Adelaide, SA, Australia

4 School of Agriculture, Food and Wine, Waite Research Institute, The University of Adelaide, Glen Osmond, SA, Australia

5 The Plant Accelerator, Australian Plant Phenomics Facility, School of Agriculture, Food and Wine, Waite Research Institute, The University of Adelaide, Glen Osmond, SA, Australia

6 Environmental Epigenetics and Genetics Group, Department of Horticulture, College of Agriculture, Food and Environment, University of Kentucky, Lexington, KY, USA

*Address all correspondence to: mjw19@aber.ac.uk
and carlos.rodriguezlopez@uky.edu

References

[1] Bird A. Perceptions of epigenetics. Nature. 2007;**447**:396-398

[2] Zilberman D, Henikoff S. Genome-wide analysis of DNA methylation patterns. Development. 2007;**134**:3959-3965

[3] Cokus SJ, Feng S, Zhang X, Chen Z, Merriman B, Haudenschild CD, et al. Shotgun bisulphite sequencing of the Arabidopsis genome reveals DNA methylation patterning. Nature. 2008;**452**:215-219

[4] Bossdorf O, Arcuri D, Richards C, Pigliucci M. Experimental alteration of DNA methylation affects the phenotypic plasticity of ecologically relevant traits in *Arabidopsis thaliana*. Evolutionary Ecology. 2010;**24**:541-553

[5] Finnegan EJ, Peacock WJ, Dennis ES. Reduced DNA methylation in *Arabidopsis thaliana* results in abnormal plant development. Proceedings of the National Academy of Sciences. 1996;**93**:8449-8454

[6] Aceituno FF, Moseyko N, Rhee SY, Gutiérrez RA. The rules of gene expression in plants: Organ identity and gene body methylation are key factors for regulation of gene expression in *Arabidopsis thaliana*. BMC Genomics. 2008;**9**:1-14

[7] Rodríguez López CM, Wetten AC, Wilkinson MJ. Progressive erosion of genetic and epigenetic variation in callus-derived cocoa (*Theobroma cacao*) plants. New Phytologist. 2010;**186**:856-868

[8] Roessler K, Takuno S, Gaut BS. CG methylation covaries with differential gene expression between Leaf and Floral Bud Tissues of *Brachypodium distachyon*. PLoS One. 2016;**11**:e0150002

[9] Widman N, Feng S, Jacobsen SE, Pellegrini M. Epigenetic differences between shoots and roots in Arabidopsis reveals tissue-specific regulation. Epigenetics. 2014;**9**:236-242

[10] Zhang M, Xu C, von Wettstein D, Liu B. Tissue-specific differences in cytosine methylation and their association with differential gene expression in sorghum. Plant Physiology. 2011;**156**:1955-1966

[11] Ay N, Janack B, Humbeck K. Epigenetic control of plant senescence and linked processes. Journal of Experimental Botany. 2014;**65**:3875-3887

[12] Xiao W, Custard KD, Brown RC, Lemmon BE, Harada JJ, Goldberg RB, et al. DNA methylation is critical for Arabidopsis embryogenesis and seed viability. The Plant Cell. 2006;**18**:805-814

[13] Cocciolone SM, Chopra S, Flint-Garcia SA, McMullen MD, Peterson T. Tissue-specific patterns of a maize Myb transcription factor are epigenetically regulated. The Plant Journal. 2001;**27**:467-478

[14] Song Q-X, Lu X, Li Q-T, Chen H, Hu X-Y, Ma B, et al. Genome-wide analysis of DNA methylation in soybean. Molecular Plant. 2013;**6**:1961-1974

[15] Lafon-Placette C, Faivre-Rampant P, Delaunay A, Street N, Brignolas F, Maury S. Methylome of DNase I sensitive chromatin in *Populus trichocarpa* shoot apical meristematic cells: A simplified approach revealing characteristics of gene-body DNA methylation in open chromatin state. New Phytologist. 2013;**197**:416-430

[16] Batut P, Dobin A, Plessy C, Carninci P, Gingeras TR. High-fidelity promoter profiling reveals widespread alternative promoter usage and transposon-driven developmental

gene expression. Genome Research. 2013;**23**:169-180

[17] Maunakea AK, Nagarajan RP, Bilenky M, Ballinger TJ, D/'Souza C, Fouse SD, et al. Conserved role of intragenic DNA methylation in regulating alternative promoters. Nature. 2010;**466**:253-257

[18] Sørensen MB, Müller M, Skerritt J, Simpson D. Hordein promoter methylation and transcriptional activity in wild-type and mutant barley endosperm. Molecular and General Genetics. 1996;**250**:750-760

[19] Zhang X, Yazaki J, Sundaresan A, Cokus S, Chan SWL, Chen H, et al. Genome-wide high-resolution mapping and functional analysis of DNA methylation in Arabidopsis. Cell. 2006;**126**:1189-1201

[20] Illingworth R, Kerr A, DeSousa D, Jørgensen H, Ellis P, Stalker J, et al. A novel CpG island set identifies tissue-specific methylation at developmental gene loci. PLoS Biology. 2008;**6**:e22

[21] Hsieh T-F, Ibarra CA, Silva P, Zemach A, Eshed-Williams L, Fischer RL, et al. Genome-wide demethylation of Arabidopsis endosperm. Science. 2009;**324**:1451-1454

[22] Ibarra CA, Feng X, Schoft VK, Hsieh T-F, Uzawa R, Rodrigues JA, et al. Active DNA demethylation in plant companion cells reinforces transposon methylation in gametes. Science. 2012;**337**:1360-1364

[23] Zemach A, Kim MY, Silva P, Rodrigues JA, Dotson B, Brooks MD, et al. Local DNA hypomethylation activates genes in rice endosperm. Proceedings of the National Academy of Sciences. 2010;**107**:18729-18734

[24] Eichten SR, Vaughn MW, Hermanson PJ, Springer NM. Variation in DNA methylation patterns is more common among maize inbreds than among tissues. The Plant Genome. 2013;**6**:1-10

[25] Mayer KFX, Waugh R, Brown JWS, Schulman A, Langridge P, et al. A physical, genetic and functional sequence assembly of the barley genome. Nature. 2012;**491**:711-716

[26] Berger B, Regt B, Tester M. Trait dissection of salinity tolerance with plant phenomics. In: Shabala S, Cuin TA, editors. Plant Salt Tolerance. Vol. 913. Humana Press; 2012. pp. 399-413

[27] Zadoks JC, Chang TT, Konzak CF. A decimal code for the growth stages of cereals. Weed Research. 1974;**14**:415-421

[28] Kitimu SR, Taylor J, March TJ, Tairo F, Wilkinson MJ, Rodriguez Lopez CM. Meristem micropropagation of cassava (*Manihot esculenta*) evokes genome-wide changes in DNA methylation. Frontiers in Plant Science. 2015;**6**:1-12

[29] Enot DP, Lin W, Beckmann M, Parker D, Overy DP, Draper J. Preprocessing, classification modeling and feature selection using flow injection electrospray mass spectrometry metabolite fingerprint data. Nature Protocols. 2008;**3**:446-470

[30] Mahalanobis PC. On the Generalised Distance in Statistics. India: Proceedings National Institute of Science; 1936. pp. 49-55

[31] Mayne BT, Leemaqz SY, Buckberry S, Rodriguez Lopez CM, Roberts CT, et al. msgbsR: An R package for analysing methylation-sensitive restriction enzyme sequencing data. Scientific Reports. 2018;**8**:2190

[32] Benjamini Y, Hochberg Y. Controlling the false discovery rate: A practical and powerful approach

to multiple testing. Journal of the Royal Statistical Society. Series B (Methodological). 1995:289-300

[33] Dunn OJ. Multiple comparisons among means. Journal of the American Statistical Association. 1961;**56**:52-64

[34] Rodríguez López CM, Morán P, Lago F, Espiñeira M, Beckmann M, Consuegra S. Detection and quantification of tissue of origin in salmon and veal products using methylation sensitive AFLPs. Food Chemistry. 2012;**131**:1493-1498

[35] Quinlan AR, Hall IM. BEDTools: A flexible suite of utilities for comparing genomic features. Bioinformatics. 2010;**26**:841-842

[36] Brockman W, Alvarez P, Young S, Garber M, Giannoukos G, Lee WL, et al. Quality scores and SNP detection in sequencing-by-synthesis systems. Genome Research. 2008;**18**:763-770

[37] Osabe K, Clement JD, Bedon F, Pettolino FA, Ziolkowski L, Llewellyn DJ, et al. Genetic and DNA methylation changes in cotton (Gossypium) genotypes and tissues. PLoS One. 2014;**9**:e86049

[38] Ichiyanagi K. Inhibition of MspI cleavage activity by hydroxymethylation of the CpG site: A concern for DNA modification studies using restriction endonucleases. Epigenetics. 2012;7(2):131-136. DOI: 10.4161/epi.7.2.18909

[39] Zhong S, Fei Z, Chen Y-R, Zheng Y, Huang M, Vrebalov J, et al. Single-base resolution methylomes of tomato fruit development reveal epigenome modifications associated with ripening. Nature Biotechnology. 2013;**31**:154-159

[40] Candaele J, Demuynck K, Mosoti D, Beemster GTS, Inze D, Nelissen H. Differential methylation during maize leaf growth targets developmentally regulated genes. Plant Physiology. 2014;**164**:1350-1364

[41] Brandeis M, Ariel M, Cedar H. Dynamics of DNA methylation during development. BioEssays. 1993;**15**:709-713

[42] Zhang X. The epigenetic landscape of plants. Science. 2008;**320**:489-492

[43] Wicker T et al. A whole-genome snapshot of 454 sequences exposes the composition of the barley genome and provides evidence for parallel evolution of genome size in wheat and barley. Plant Journal. 2009;**59**:712-722

[44] Bewick AJ, Ji L, Niederhuth CE, Willing E-M, Hofmeister BT, Shi X, et al. On the origin and evolutionary consequences of gene body DNA methylation. Proceedings of the National Academy of Sciences. 2016;**113**:9111-9116

[45] Deaton AM, Bird A. CpG islands and the regulation of transcription. Genes & Development. 2011;**25**:1010-1022

[46] Xie H, Konate M, Sai N, Tesfamicael KG, Cavagnaro T, Gilliham M, et al. Global DNA methylation patterns can play a role in defining terroir in grapevine (*Vitis vinifera* cv. Shiraz). Frontiers in Plant Science. 2017;**8**:1860

[47] Li-Byarlay H, Li Y, Stroud H, Feng S, Newman TC, Kaneda M, et al. RNA interference knockdown of DNA methyl-transferase 3 affects gene alternative splicing in the honey bee. Proceedings of the National Academy of Sciences. 2013;**110**:12750-12755

[48] Jones PA. Functions of DNA methylation: Islands, start sites, gene bodies and beyond. Nature Reviews Genetics. 2012;**13**:484-492

[49] Hollister JD, Gaut BS. Epigenetic silencing of transposable elements:

A trade-off between reduced transposition and deleterious effects on neighboring gene expression. Genome Research. 2009;**19**:1419-1428

[50] Druka A, Muehlbauer G, Druka I, Caldo R, Baumann U, Rostoks N, et al. An atlas of gene expression from seed to seed through barley development. Functional & Integrative Genomics. 2006;**6**:202-211

[51] Orgel LE, Crick FHC. Selfish DNA: The ultimate parasite. Nature. 1980;**284**:604-607

[52] Lister R, O'Malley RC, Tonti-Filippini J, Gregory BD, Berry CC, Millar AH, et al. Highly integrated single-base resolution maps of the epigenome in Arabidopsis. Cell. 2008;**133**:523-536

[53] Hirsch CD, Springer NM. Transposable element influences on gene expression in plants. Biochimica et Biophysica Acta—Gene Regulatory Mechanisms. 1860;**2016**:157-165

[54] Marquez Y, Brown JWS, Simpson C, Barta A, Kalyna M. Transcriptome survey reveals increased complexity of the alternative splicing landscape in Arabidopsis. Genome Research. 2012;**22**:1184-1195

[55] Simpson CG, Lewandowska D, Fuller J, Maronova M, Kalyna M, Davidson D, et al. Alternative splicing in plants. Biochemical Society Transactions. 2008;**36**:508-510

[56] Barbazuk WB, Fu Y, McGinnis KM. Genome-wide analyses of alternative splicing in plants: Opportunities and challenges. Genome Research. 2008;**18**:1381-1392

[57] Warf MB, Berglund JA. The role of RNA structure in regulating pre-mRNA splicing. Trends in Biochemical Sciences. 2010;**35**:169-178

[58] Pan Q, Shai O, Lee LJ, Frey BJ, Blencowe BJ. Deep surveying of alternative splicing complexity in the human transcriptome by high-throughput sequencing. Nature Genetics. 2008;**40**:1413-1415

[59] Lorincz MC, Dickerson DR, Schmitt M, Groudine M. Intragenic DNA methylation alters chromatin structure and elongation efficiency in mammalian cells. Nature Structural Molecular Biology. 2004;**11**:1068-1075

Chapter 8

Library Preparation for Whole Genome Bisulfite Sequencing of Plant Genomes

Kendall R. Corbin and Carlos M. Rodriguez Lopez

Abstract

Epigenetic mechanisms are a key interface between the environment and the genotype. These mechanisms regulate gene expression in response to plant development and environmental stimuli, which ultimately affects the plant's phenotype. DNA methylation, in particular cytosine methylation, is probably the best studied epigenetic modification in eukaryotes. It has been associated to the regulation of gene expression in response to cell/tissue differentiation, organism development and adaptation to changing environments. Whole genome bisulfite sequencing (WGBS) is considered the gold standard to study DNA methylation at a genome level. Here we present a protocol for the preparation of whole genome bisulfite sequencing libraries from plant samples (grapevine leaves) which includes detailed instructions for sample collection and DNA extraction, sequencing library preparation and bisulfite treatment.

Keywords: whole genome bisulfite sequencing, methylome analysis, DNA methylation, epigenetic modifications, *Vitis vinifera*

1. Introduction

Plants being sessile have developed strategies to adapt to their environment, specifically via epigenetic modification of their genome [1, 2]. Epigenetic mechanisms, both heritable and reversible, allow an organism to respond to its environment through changes in gene expression, without changing the underlying genome [3–6]. One of the most widely studied epigenetic mechanisms is cytosine methylation (5mC), which is the result of a methyl group replacing a hydrogen in the cyclic carbon-5 of cytosines. In plants, methylation of cytosine bases can occur in three contexts (DNA base sequences) CG, CHG or CHH, where H is any nucleotide other than G [7]. Plant nuclear genomes are known to contain more extensive and expansive DNA methylation than that found in animals [8]. DNA methylation has been identified in a range of plants and plays a role in a wide variety of biological processes from plant development and organ differentiation to response to stress [9–20].

Due to the functional importance of DNA methylation in many species, a plethora of DNA methylation analysis approaches has been developed in recent years. These can be mainly grouped into three functional types that (1) indicate the methylation status of a specific sequence; (2) reveal the degree and patterning of DNA methylation across partly characterized genomes; or (3) facilitate the discovery and sequencing of new epialleles [7]. From a technical point of view, such methodologies can be grouped into those using global estimation of all nucleic base species (e.g. HPLC and LC-MS/MS), methylation-sensitive restriction enzymes [18, 21, 22], high-resolution melting analysis [10, 23, 24], methylcytosine-specific antibodies and methylated DNA-binding domains [25, 26], bisulfite conversion of DNA, and third-generation DNA sequencing technologies, including single molecule real-time (SMRT) sequencing and nanopore sequencing (for extensive reviews in these methodologies, see [27–29]).

Of all these techniques currently available, only bisulfite conversion of DNA and third-generation DNA sequencing provide a single-base resolution view of methylated cytosines across the selected target sequence. This approach is not limited by genome size and may be applied to a relatively small fraction of a genome or a whole genome. More recently developed techniques are capable of reading 5mC, and other DNA modifications, without the need for any chemical alteration of the target DNA molecule. However, their throughput, accuracy and affordability are still not sufficient for routine use. Bisulfite conversion of DNA, in turn, is based on the selective chemical modification of unmethylated cytosines (C) into uracils (U) (which are read as thymines (T) by DNA polymerases during PCR amplification) (**Figure 1**), while leaving unchanged 5mC (**Figure 2**). Due to its high throughput, reliability and low cost, bisulfite conversion is considered the "gold standard" DNA methylation analysis. Next-generation sequencing (NGS) allows the rapid sequencing of whole genomes. Combined with bisulfite conversion of the target DNA, it also permits the identification of methylated cytosines at a single-base resolution of whole genomes (i.e. whole genome bisulfite sequencing (WGBS)).

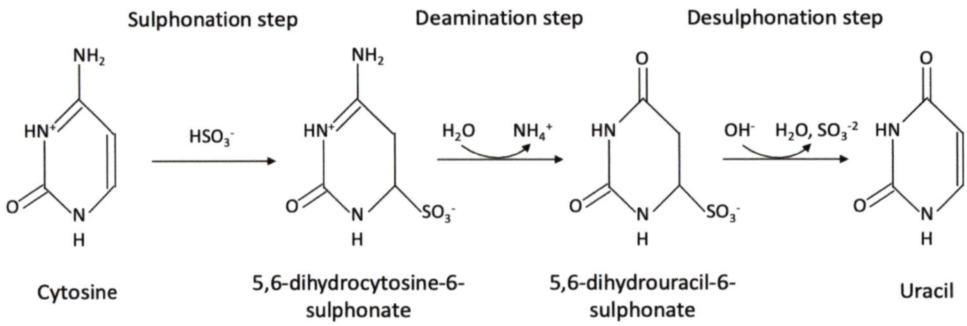

Figure 1.
Bisulfite conversion of unmethylated cytosines. Bisulfite conversion reaction starts with the addition of a sodium bisulfite group (sulphonation step) to the pyrimidine ring double bond between carbons 5 and 6 to form a 5,6-dihydrocytosine-6-sulphonate. Next, spontaneous and irreversible hydrolytic deamination results in a 5,6-dihydrouracil-6-sulphonate (deamination step). Finally, high pH conditions favor the loss of the sulphonate group (desulphonation step) to form uracil. Only unmethylated cytosines are susceptible to the bisulfite reaction. Methylated (5mC and 5-hmC) cytosines do not undergo conversion.

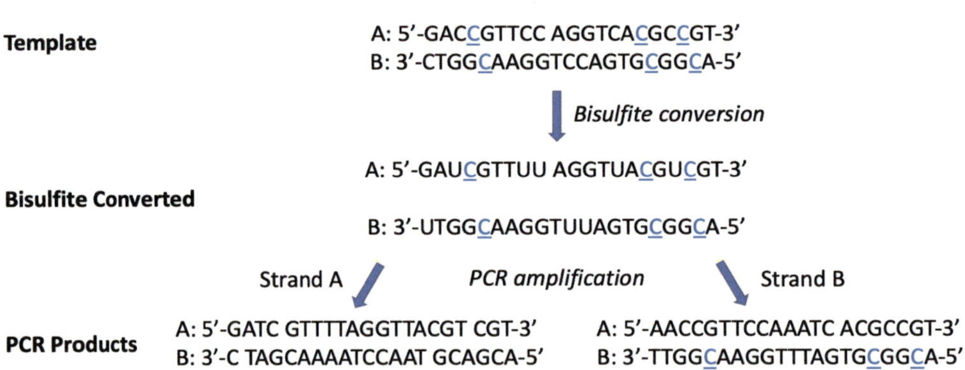

Figure 2.
Bisulfite conversion of a sample DNA sequence. Nucleotides highlighted in blue (methylated cytosines) are protected from bisulfite conversion and are maintained as cytosines. Unmethylated cytosines are converted to uracils. Loss of the original base pairing will yield two different PCR products from each DNA fragment.

2. Procedure

2.1 Equipment

i. Ultralow freezer (−80°C)

ii. Mortar and pestle—used to grind leaf samples prior to DNA extraction. Use a clean set for each sample to avoid cross-contamination. Wash both parts using warm water and soap, air-dry, wrap in aluminium foil and autoclave.

iii. NanoDrop™ 2000 Spectrophotometer—this UV-Vis spectrophotometer has the capability to quantify and assess the purity of small volumes of DNA (0.5 μL). The sample may be pipetted directly onto the optical measurement surface. Additional information regarding DNA quantification and quality assessment using NanoDrop can be found at https://www.thermofisher.com/us/en/home/industrial/spectroscopy-elemental-isotope-analysis/molecular-spectroscopy/ultraviolet-visible-visible-spectrophotometry-uv-vis-vis/uv-vis-vis-instruments/nanodrop-microvolume-spectrophotometers/nanodrop-nucleic-acid-quantification.html

iv. Thermocycler (PCR machine)—thermocyclers amplify segments of nucleic acid following a series of temperature-controlled enzymatic reactions.

v. Covaris M220 Focused-Ultrasonicator™ and MicroTUBE-50 (Covaris, catalog number: 520166) (or equivalent models and parts)—sonicators are used for shearing DNA to a desired size.

vi. Magnetic rack for 1.5 mL tubes—magnetic racks are used for separation and purification of nucleic acids in combination with paramagnetic beads (e.g. AMPure XP beads).

vii. Qubit Fluorometric Quantification and Qubit dsDNA HS (High Sensitivity) Assay Kit (Thermo Fisher Scientific, catalog number: Q32854). Qubit assays accurately quantify nucleic acids quickly and require small volumes of sample.

viii. Agilent Fragment Analyzer, Agilent Bioanalyzer (Agilent Technologies) or the Bio-Rad Experion (Bio-Rad Laboratories).

ix. High-speed centrifuge.

2.2 Consumables

i. Sterile microcentrifuge tubes 1.5 mL (Eppendorf® Safe-Lock™)

ii. 15 mL polypropylene centrifuge tubes (Laboratory Product Sales)

iii. Filtered pipette tips

iv. Wide-bore pipette tips

v. Sterile 200 μL PCR tubes

vi. Sterile 500 μL tubes

2.3 Chemicals and reagents

i. Molecular biology grade ethanol (MilliporeSigma or Fisher BioReagents)

ii. Molecular biology grade water (MilliporeSigma)

iii. Cetyltrimethylammonium bromide (CTAB)

iv. Ethylenediaminetetraacetic acid (EDTA)

v. Tris hydrochloride (Tris-HCl)

vi. Hydrochloric acid (HCl)

vii. Polyvinylpolypyrrolidone (PVP)

viii. Chloroform

ix. Octane

x. Sodium chloride (NaCl)

xi. RNAse A (Sigma-Aldrich, catalog number: R4642)

xii. Agencourt AMPure XP magnetic beads (Beckman Coulter, catalog number: A63880)

xiii. Q5® High-Fidelity 2× Master Mix (New England Biolabs, catalog number: M0492S)

xiv. 10× End Repair Buffer (New England Biolabs, catalog number: B6052S)

xv. End Repair Enzyme Mix (New England Biolabs, catalog number: E6051)

xvi. 10× dA-Tailing Reaction Buffer (New England Biolabs, catalog number: B6059S)

xvii. A-tailing Enzyme (e.g. Klenow Fragment (3′ → 5′ exo-) (New England Biolabs, catalog number: M0212S)

xviii. 10× T4 DNA Ligation Buffer and T4 DNA Ligase (New England Biolabs, catalog numbers: B0202S and M0202)

xix. TruSeq Sequencing adapters: adapters are ordered as lyophilized oligonucleotides with the specified modifications[1] from the provider of

[1] Order the oligonucleotides with standard desalting. Request that all cytosines are methylated. This will allow the sequence integrity of the adapters to be maintained after bisulfite treatment. Also, order the indexed adapter with a 5′ phosphate group and TruSeq Universal Adapter with phosphorothioate bond between the 3′ end C and T nucleotides.

your choice (sequences are provided below[2]). To prepare the adapters, resuspend both oligonucleotides with TE buffer to a final concentration of 200 μM. Then add 75 μL from each into a 200 μL sterile PCR tube. To allow annealing of the complementary sections of the oligos, heat the mixture using a thermocycler to 95°C for 1 min, and then slowly lower the temperature to 30°C at a rate of 1°C/min. This can be accomplished by programming your thermocycler with a single step PCR cycle at 95°C for 1 min followed by 65 cycles during which the temperature is reduced by 1°C each cycle. Store double-stranded adapters at −20°C.

I. TruSeq universal adapter: 5′-AATGATACGGCGACCACCGAGATCTACACTCTTTCCCTACACGACGCTCTTCCGATC*T-3′

II. TruSeq INDEX adapter: 5′-P*GATCGGAAGAGCACACGTCTGAACTCCAGTCAC[i7]ATCTCGTATGCCGTCTTCTGCTTG-3′

xx. Library amplification primers: primers are ordered as lyophilized oligonucleotides with standard desalting from the provider of your choice (sequences are provided below). To prepare the primers, resuspend both oligonucleotides with TE buffer to a final concentration of 100 μM. (This is your stock solution. Store at −20°C.) To prepare the Forward and Reverse Primer Mix, mix 10 μL from each in a new tube, and add 80 μL of molecular grade water to achieve a final concentration of 10 μM.

I. Forward primer: 5′-AATGATACGGCGACCACCGAGATCTACACTCTTTCCCTACACGA-3′

II. Reverse primer: 5′-CAAGCAGAAGACGGCATACGAGAT-3′

2.4 Additional items required

i. Insulated polystyrene box

ii. Pipettes

iii. Water bath

iv. Liquid nitrogen

v. Refrigerator (4°C) and freezer (−20°C)

3. Set-up

i. Label all tubes prior to starting any of the described protocols to reduce the likeliness of downstream errors.

---

[2] [i7] index sequences can be found at https://support.illumina.com/content/dam/illumina-support/documents/documentation/chemistry_documentation/experiment-design/illumina-adapter-sequences-1000000002694-09.pdf

ii. Use sterilized tools (scissors, knives, tweezers, etc.) for harvesting plant material, and clean utensils thoroughly between samples using 70% (v/v) ethanol.

iii. Gloves should be worn at all times while handling samples to minimize cross-contamination (change gloves as needed).

iv. DNA extractions, next-generation sequencing library preparations and bisulfite treatments should be carried out in a PCR cabinet or similar to minimize contamination.

v. General safety notes.

- Follow safe operating procedures when handling cryogenic products (dry ice and liquid nitrogen). Prior to usage (and transport) of cryogenic products, a risk assessment should be conducted to evaluate hazards and identify control measures that may be implemented to minimize the level of risk. Additional information about cryogenic materials precautions and safe handling procedures may be available from your local Office of Environmental Health and Safety.

- β-Mercaptoethanol (also known as 2-hydroxyethylmercaptan, BME or thioethylene glycol) is a toxic chemical that should be handled with extreme caution. Exposure to this product may cause respiratory issues, vomiting or skin irritation. Long-term exposure to this product can result in death. Personal protective equipment should be worn when handling this product and all experimental work conducted in a fume hood. Hazard control measures include wearing nitrile laboratory gloves (if gloves get splashed or tear, change immediately), safety glasses, closed toe shoes, a laboratory coat, and if spills are possible, a face shield. Safety documentation about this product, including information relevant to storage, transport and disposal, may be found on manufacturers Website.

3.1 Collection of plant material

i. Collect three individual leaves at bud burst (E-L 7 [30]) from the number of desired grapevines. The rationale for using immature vegetative tissue (leaves) is that cell number is fixed very early during development; thus the number of genome copies per gram of tissue is higher in younger leaves relative to older leaves. It is also advantageous to use younger plant material as some plant species accumulate secondary metabolites (such as alkaloids and flavonoids) as their tissues age. High levels of these metabolites can impede DNA extraction or PCR amplification [31].

Note: DNA methylation has been shown to change with the plant's circadian cycle [32] and during plant development [19]. Thus, when collecting samples for DNA methylation analysis from more than one plant, it is extremely important to harvest all plant tissue at approximately the same time of day and at the same developmental stage in order to minimize unwanted variability in DNA methylation.

ii. Immediately upon harvesting the leaves, put the material in a pre-labelled 1.5 mL centrifuge tube. Place the tubes in an insulated container (i.e. polystyrene box) and cover with dry ice (solid CO_2).

Note: By immediately snap-freezing the samples, changes in DNA methylation profiles induced during harvesting and cell death will be minimized.

iii. Store all samples at −80°C until required for DNA extraction.

Note: Storage of samples at ultralow temperatures will minimize DNA degradation. Avoid unnecessary freeze-thawing cycles, including during the period of material transport from the field to laboratory.

3.2 Recipes for buffers, solutions and reagents

i. Ethanol (70, 80 and 95% v/v). Store at room temperature.

ii. CTAB DNA extraction buffer (per 100 mL): 20 mM sodium EDTA (1 mL of 0.5 M stock) and 100 mM Tris-HCl (10 mL of 1 M stock), adjust pH to 8.0 with HCl; add 1.4 M NaCl (8.2 g), 1% (w/v) PVP (1.0 g), and 2.0% (w/v) CTAB (2.0 g). Dissolve CTAB by heating to 60°C. Store at 37°C.

iii. Chloroform-octanol 24:1 (v/v). Store at room temperature.

iv. 5 M sodium chloride (NaCl)—dissolve 292 g of NaCl in 800 mL of water, and then adjust the volume to 1 L with water.

v. 1× Tris-EDTA buffer (TE buffer)—10 mM Tris-HCl and 1 mM EDTA, adjust pH to 8.0 and autoclave. Store at room temperature

4. Protocol

4.1 DNA extraction

DNA extraction is carried out following a modified CTAB protocol [33].

i. Pour liquid nitrogen on to a mortar and pestle.

Note: The mortar should be fully cooled in liquid nitrogen prior to and during usage. In addition, the sample must remain frozen during the grinding process. Accidental thawing may result in DNA degradation.

ii. Grind 500 mg of leaf material in a mortar and pestle. Continue to add liquid nitrogen to ensure the equipment remains cold.

Note: Over grinding of plant biomass will cause DNA shearing, which results in lower yields after bisulfite treatment due to degradation of small DNA fragments.

iii. Add 5 mL of CTAB extraction buffer to the ground leaves and mix with a sterile spatula.

iv. Transfer the slurry to a 15 mL polypropylene centrifuge tube. Rinse the mortar and pestle with 1 mL of extraction buffer, and add to the tube (added to original extract).

v. Add 50 mg polyvinylpolypyrrolidone (PVP), screw the cap on the tube tightly, and invert the tube several times to mix thoroughly.

Note: PVP is added at a concentration of 100 mg PVP/g leaf tissue used in step ii.

vi. Incubate the tube in a water bath set at 60°C for 25 min. Carefully remove the tube from the bath and cool to room temperature.

Note: Take care when removing the sample from the water bath, wear personal protective equipment (laboratory jacket, safety glasses and heat-resistant gloves).

vii. Centrifuge the homogenate for 5 min at 14,000 × *g* (room temperature), and transfer the supernatant to a clean 1.5 mL tube.

viii. Treat with 1 μL RNase A per 100 μL DNA solution and incubate at 37°C for 15 min.

Note: An RNAse treatment step is included to enzymatically digest RNA in the material, minimizing the amount of RNA extracted with the DNA. Contaminating RNA will result in the overestimation of DNA quantity.

ix. Add 6 mL of chloroform-octanol, and mix gently by inverting the tube 20–25 times to form an emulsion.

x. Spin at 14,000 × *g* for 15 min in a centrifuge (room temperature).

xi. Using a wide-bore pipette tip, transfer the top aqueous phase to a new 15 mL tube. A second chloroform-octanol extraction may be performed if the aqueous phase is cloudy due to the presence of PVP (repeat steps ix to xi).

xii. Add 3 mL of 5 M NaCl to the aqueous solution and mix well (invert gently by hand).

xiii. Add two volumes of cold (−20°C) 95% (v/v) ethanol and refrigerate (4–6°C) until DNA strands begin to appear.

Note: The solution should be left for at least 15 min but can stay refrigerated for longer if necessary.

xiv. Spin at 10,000 × *g* for 3 min (room temperature).

xv. Increase the speed of the centrifuge to 14,000 × *g*. Spin samples for an additional 3 min.

Note: Differential centrifugation steps aid in keeping the DNA at the bottom of the tube.

xvi. Carefully pour off supernatant and wash pellet with 1 mL of chilled (0–4°C) 70% (v/v) ethanol.

xvii. Remove ethanol by pipetting—do not disturb the DNA pellet. Air-dry the remaining ethanol by leaving the tubes uncovered at room temperature for 10 min.

xviii. Solubilize the DNA pellet in 200–300 μL TE buffer.

xix. Quantify isolated DNA using the NanoDrop™ 2000.

Note: TE buffer should be used as the reference blank.

xx. Normalize DNA concentrations to 20 ng/μL using molecular grade water.

xxi. Store DNA samples at −20°C (short-term) or −80°C (long-term).

4.2 DNA shearing

i. Aliquot 1 μg of genomic DNA (equivalent to 50 μL of DNA with a concentration of 20 ng/μL) into a Covaris MicroTUBE-50, and add 5 μL of molecular biology water. The final volume in the microtube is 55 μL.

ii. Shear DNA to 200 bp fragments using the Covaris M220 Focused-Ultrasonicator™, using the following specifications:

Duration, 90 s; peak power, 75 W; duty factor, 25%; cycles per burst, 1000

iii. Transfer 50 μL of the fragmented DNA to a clean, pre-labelled 200 μL PCR tube.

Note: Label the top and the side of the PCR tubes.

4.3 Sheared DNA end repair

i. Prepare End Repair Master Mix containing 8 μL molecular grade water, 7 μL of 10× end repair buffer and 5 μL end repair enzyme.

Note: When preparing Master Mixes, prepare 10% extra to account for pipetting errors, and allow enough reaction mix for all sample. For example, for 10 samples, prepare enough Master Mix for those samples plus one extra (11 in total): combine 88 μL molecular grade water, 77 μL of 10× end repair buffer and 55 μL end repair enzyme.

ii. Add 20 μL of End Repair Master Mix to each of the sheared samples.

iii. Incubate in a thermocycler at 20°C for 30 min.

Note: At this point remove AMPure XP beads from the refrigerator and allow the bottle to reach room temperature before use. Immediately before pipetting, resuspend the beads by vortexing vigorously. The AMPure purification system selectively binds DNA fragments to paramagnetic beads, allowing the removal of excess primers, nucleotides, salts and enzymes during a simple washing step. These clean-up steps result in a more purified PCR product. For further information about using AMPure XP for PCR purification, please refer to the manufacturer's manual.

iv. Capture DNA by adding 120 μL of AMPure XP beads, pipette up and down to achieve a homogenous mixture, and incubate at room temperature for 5 min.

v. Transfer the beads with captured DNA to a 1.5 mL tube.

vi. Place the tube on a magnetic rack for 2 min.

vii. Keep the tube on the magnetic rack and remove the supernatant using a pipette. Do not disturb the beads.

Note: The aim of this step is to remove the AMPure XP buffer. At this stage the DNA is captured by the beads which are kept in the tube by the magnet. The buffer can be discarded.

viii. Keep the tube on the magnetic rack and add 200 μL of 80% (v/v) ethanol.

Note: Due to the different evaporation rates of H_20 and ethanol, it is important to use freshly prepared ethanol.

ix. Incubate for 30 s on the magnetic rack and use a pipette to remove the ethanol.

x. Repeat steps viii and ix.

Note: After the second ethanol wash, remove as much ethanol as possible using a 10 μL pipette. These wash steps are important to remove any remains of the End Repair Master Mix. At this stage the DNA is captured by the AMPure beads which are kept in the tube by the magnet.

xi. Remove residual ethanol by leaving the tube open on the magnetic rack for 5 min (air-dry).

Note: Do not over dry the beads as it will lower DNA yields. Appearance of cracks on the bead pellet is indicative of over drying.

xii. Remove the tube from the magnetic rack, add 42 μL of molecular grade water, and pipette up and until beads are fully resuspended.

xiii. Leave the tube at room temperature for 5 min to allow the DNA to be released from the AMPure beads.

xiv. Place the tube in the magnetic rack and leave at room temperature for 2 min.

xv. Transfer 40 μL of the supernatant to a clean 200 μL PCR tube.

Note: At this stage the DNA is resuspended in the water. Beads can be safely discarded. Do not attempt to pipette the entire volume in the tube (42 μL) as some of the AMPure beads may be transferred which could affect later reactions. If beads are disturbed during pipetting, simply put the whole volume back in the tube and proceed from step xiv.

4.4 Fragmented DNA A-tailing

i. Prepare the A-tailing Master Mix containing 2 μL molecular grade water, 5 μL of 10× A-tailing buffer and 3 μL A-tailing enzyme.

Note: When preparing Master Mixes, prepare 10% extra to account for pipetting errors and allow enough reaction mix for all samples.

ii. Add 10 μL of A-tailing Master Mix to each of the samples (200 μL PCR tube).

iii. Incubate in a thermocycler at 30°C for 30 min.

iv. Capture DNA by adding 90 μL of AMPure XP beads, pipette up and down to achieve a homogenous mix, and leave at room temperature for 5 min.

v. Transfer the beads with the capture DNA to a clean 1.5 mL tube.

vi. Place the tube on a magnetic rack for 2 min.

vii. Keep the tube on the magnetic rack and remove the supernatant without disturbing the beads using a pipette.

Note: The aim of this step is to remove the AMPure XP buffer. At this stage the DNA is captured by the beads which are kept in the tube by the magnet. The buffer can be safely discarded.

viii. Keep the tube on the magnetic rack and add 200 μL of 80% (v/v) ethanol.

ix. Incubate for 30 s on the magnetic rack and use a pipette to remove the ethanol.

x. Repeat steps viii and ix.

xi. Evaporate ethanol by leaving the tube open on the magnetic rack for 5 min.

xii. Remove the tube from the magnetic rack and resuspend the beads by adding 32 μL of molecular grade water and pipette up and down.

xiii. Leave the tube at room temperature for 5 min to allow the DNA to be released from the AMPure beads.

xiv. Place the tube in the magnetic rack and leave at room temperature for 2 min.

xv. Transfer 30 μL of the supernatant to a clean 200 μL PCR tube. Do not transfer beads.

4.5 Ligation of sequencing adapters

i. Prepare the Ligation Master Mix containing 5 μL of 10× Ligation Buffer, 2.5 μL T4 DNA Ligase and 7.5 μL molecular grade water.

ii. Add 5 μL of TruSeq Adapter to each of the samples in a 200 μL PCR tube.

Note: Add 5 μL of adapters (10 μM) for every 1 μg of starting DNA. If you are planning to multiplex more than one sample in each sequencing lane, use adapters with different index sequences.

iii. Add 15 μL of Ligation Master Mix to each of the samples in the 200 μL PCR tube and mix by pipetting up and down.

iv. Incubate in a thermocycler at 20°C for 15 min.

v. Capture DNA by adding 90 μL of AMPure XP beads, pipette up and down to achieve a homogenous mix. Leave at room temperature for 5 min.

vi. Transfer the beads with the captured DNA to a clean 1.5 mL tube.

vii. Place the tube on a magnetic rack for 2 min.

viii. Keep the tube on the magnetic rack, and remove the supernatant without disturbing the beads using a pipette.

Note: The aim of this step is to remove the AMPure XP buffer. At this stage the DNA is captured by the beads which are kept in the tube by the magnet. The buffer can be safely discarded.

ix. Keep the tube on the magnetic rack and add 200 μL of 80% (v/v) ethanol.

x. Incubate for 30 s on the magnetic rack and use a pipette to remove the ethanol.

xi. Repeat steps ix and x.

xii. Evaporate ethanol by leaving the tube open on the magnetic rack for 5 min.

xiii. Remove the tube from the magnetic rack, add 105 μL of molecular grade water, and pipette up and down until beads are resuspended.

xiv. Leave the tube at room temperature for 5 min to allow the DNA to be released from the AMPure beads.

xv. Place the tube in the magnetic rack and leave at room temperature for 2 min.

xvi. Transfer 100 μL of the supernatant to a clean 1.5 mL tube. Do not transfer beads.

4.6 Sequencing library fragment size selection

i. Add 60 μL of AMPure beads to capture DNA fragments >450 bp, pipette up and down to achieve a homogenous mix, and leave at room temperature for 5 min.

Note: Beads preferentially capture larger fragments of DNA. The size range that the beads capture is determined by the volume to volume ratio of AMPure XP buffer and DNA aqueous solution. In this case a ratio of 0.6 (60 μL AMPure XP buffer/100 μL DNA) will capture fragments above 450 bp.

ii. Place the tube on a magnetic rack for 2 min.

iii. With the tube on the magnetic rack, transfer 155 μL of supernatant to a new tube without disturbing the beads.

Note: Do not discard the supernatant in this case. The supernatant contains the fragment size range required for sequencing, while larger, unwanted fragments are still captured by the beads. At this stage the beads and the tube containing them can be discarded.

iv. Add 20 μL of beads to the 155 μL of supernatant collected in step iii, pipette up and down to achieve a homogenous mix, and leave at room temperature for 5 min.

v. Place the tube on a magnetic rack for 2 min.

vi. Keep the tube on the magnetic rack, and remove the supernatant without disturbing the beads using a pipette.

Note: In this case a ratio of 0.88 (82 μL AMPure XP buffer/93 μL DNA) will capture fragments above 100 bp. The supernatant, containing unligated TruSeq adapters or DNA fragments below that size can be safely discarded.

vii. Keep the tube on the magnetic rack and add 200 μL of 80% (v/v) ethanol.

viii. Incubate for 30 s on the magnetic rack and use a pipette to remove the ethanol.

ix. Repeat steps vii and viii.

x. Evaporate ethanol by leaving the tube open on the magnetic rack for 5 min.

xi. Remove the tube from the magnetic rack and resuspend the beads by adding 22 μL of molecular grade water and pipette up and down until beads are fully resuspended.

xii. Leave the tube at room temperature for 5 min to allow the DNA to be released from the AMPure beads.

xiii. Place the tube in the magnetic rack and leave at room temperature for 2 min.

xiv. Transfer 20 μL of the supernatant to a clean 200 μL PCR tube. Make sure not to transfer the beads.

Storage: At this stage the size-selected samples can be stored until required for bisulfite treatment. For short-term storage keep at −20°C, for long-term store at −80°C.

4.7 Bisulfite conversion of size-selected library

DNA samples are bisulfite converted using the EZ DNA Methylation-Lightning Kit (Zymo Research).

i. Thaw samples completely (if stored in the freezer prior to bisulfite treatment), and centrifuge to bring droplets to the bottom.

ii. Add 130 μL of Lightning Conversion Reagent to the tube containing the 20 μL size-selected library.

Note: Mix and then centrifuge briefly to ensure there are no droplets in the cap or sides of the tube.

iii. Place the PCR tube in a thermal cycler and incubate using the following programme:

a. 98°C for 8 min[3]

[3] High temperature is used to achieve complete denaturation of the double stranded DNA molecule and to favor the forward reaction during the reversible sulphonation step.

b. 54°C for 60 min[4]

c. 4°C storage for up to 20 h[5]

iv. Add 600 μL of M-Binding Buffer to a Zymo-Spin™ IC Column, and place the column into the collection tube (provided by supplier).

Note: Do not touch the bottom of the column with a pipette tip; this may damage the filtering matrix.

v. Load the sample (from step iii) into the Zymo-Spin™ IC Column containing the M-Binding Buffer. Close the cap and mix by inverting the column 10 times.

Note: Do not touch the bottom of the column with a pipette tip; this may damage the filtering matrix.

vi. Centrifuge at full speed (>10,000 × *g*) for 30 s. Discard the flow-through.

Note: At this stage the DNA is captured in the column matrix and the flow-through liquid can be safely discarded.

vii. Add 100 μL of M-Wash Buffer[6] to the column. Centrifuge at full speed (>10,000 × *g*) for 30 s in benchtop centrifuge. Discard the flow-through.

Note: This is a wash step. At this stage, the DNA is still captured in the column matrix and the flow-through can be safely discarded.

viii. Add 200 μL of L-desulphonation buffer to the column, and leave at room temperature (20–30°C) for 15–20 min.

Note: This is an alkali desulphonation step that chemically removes the $SO3^2$ group added to unmethylated cytosines during the sulphonation step (**Figure 1**). At the end of this stage, cytosines that were originally unmethylated will be converted to uracils.

ix. After the incubation period, centrifuge at full speed for 30 s. Discard the flow-through.

Note: The aim of this centrifugation step is to remove the L-desulphonation buffer. At this stage the DNA is still captured in the column matrix.

x. Add 200 μL of M-Wash Buffer to the column. Centrifuge at full speed for 30 s. Discard the flow-through.

---

[4] This step consists of two consecutive chemical reactions. First, a sulphonation step selectively adds a SO_3^- group to unmethylated cytosines leaving methylated cytosines unchanged. Then, a spontaneous hydrolytic deamination exchanges de amino group (NH_2) for an oxygen atom in the sulfonated cytosines during the sulphonation step (**Figure 1**).

[5] The 4°C storage step is optional. Ideally continue with the rest of the protocol right after the incubation. Longer storage at 4°C could result in DNA degradation.

[6] Ensure that molecular grade 100% ethanol is added to the M-DNA Wash Buffer as recommended by the manufacturer. For example, add 24 mL of ethanol to the 6 mL M-Wash Buffer concentrate (D5030) or 96 mL to the 24 mL M-Wash Buffer concentrate (D5031). M-DNA Wash Buffer included with D5030S and D5030T kits is supplied ready-to-use and does not require the addition of ethanol.

xi. Add 200 μL of M-Wash Buffer to the column. Centrifuge at full speed for 30 s. Discard the flow-through and collection tube. Keep the column matrix.

Note: These are wash steps. At this stage the DNA is still captured in the column matrix, and the flow-through can be safely discarded.

xii. Place the column into a 1.5 mL microcentrifuge tube, and add 12 μL of M-Elution Buffer directly to the column matrix. Centrifuge for 30 s at full speed to elute the DNA.

Storage: Ideally use bisulfite-treated DNA immediately after treatment. After bisulfite conversion of non-methylated cytosines into uracils, genomic DNA does not maintain its original base pairing. This typically leads to single-stranded A-, U-, and T-rich DNA that is more susceptible to degradation. Long-term storage of bisulfite-converted DNA will lead to loss of sample concentration. If long-term storage is required, place in an ultralow freezer (−80°C).

4.8 PCR amplification of bisulfite-converted library

i. Prepare the PCR Master Mix: 25 μL Q5® High-Fidelity 2× Master Mix, 2.5 μL Forward and Reverse Library Amplification Primer Mix at 10 μM and 12.5 μL molecular grade water.

Note: When preparing Master Mixes, prepare 10% extra to account for pipetting errors and allow enough reaction mix for all samples.

ii. Thaw samples completely (if stored prior to bisulfite treatment) and centrifuge to bring droplets to the bottom.

iii. Transfer 10 μL of the bisulfite-treated library to a new 200 μL PCR tube.

iv. Add 40 μL of PCR Master Mix to each tube.

v. Place the PCR tube/tubes in a thermal cycler and incubate using the following program:

98°C for 30 s
98°C for 30 s
60°C for 30 s
Go to step 2: 7–12 times[7]
72°C for 4 min
72°C for 10 min
4°C hold[8]

vi. Centrifuge the PCR tube for a few seconds to ensure there are no droplets in the cap or sides of the tube due to condensation generated during PCR amplification.

[7] Maintain the number of cycles as low as possible to minimize DNA polymerase base substitution errors.

[8] After PCR amplification, bisulfite-treated DNA recovers its base pairing. This stabilizes the DNA molecule making long-term storage possible.

vii. Add 45μL of beads to the PCR product, pipette up and down to achieve a homogenous mix, and leave at room temperature for 5 min.

viii. Place the tube on a magnetic rack for 2 min.

ix. Keep the tube on the magnetic rack, and remove the supernatant without disturbing the beads using a pipette.

Note: In this case a ratio of 0.9 (45 μL AMPure XP buffer/50 μL PCR product) will capture fragments above 100 bp. The supernatant containing unused PCR primers or DNA fragments below that size can be safely discarded.

x. Keep the tube on the magnetic rack and add 200 μL of 80% (v/v) ethanol.

xi. Incubate for 30 s on the magnetic rack and use a pipette to remove the ethanol.

xii. Repeat steps ix and x.

xiii. Air-dry any ethanol by leaving the tube open on the magnetic rack for 5 min.

xiv. Remove the tube from the magnetic rack and resuspend the beads by adding 22 μL of molecular grade water and pipette up and down until beads are fully resuspended.

xv. Leave the tube at room temperature for 5 min to allow the DNA to be released from the AMPure beads.

xvi. Place the tube in the magnetic rack and leave at room temperature for 2 min.

xvii. Transfer 20 μL of the supernatant to a clean 500 μL tube. Make sure not to transfer the beads.

xviii. Check sequencing library concentration using Qubit and fragment size distribution using the Agilent Fragment Analyzer, Agilent Bioanalyzer (Agilent Technologies) or the Bio-Rad Experion (Bio-Rad).

Note: A good WGBS library should show a fragment distribution between 150 and 500 bp (**Figure 3** Box B). Smaller peaks in the electropherogram would be indicative of sequencing adapters or PCR primers (**Figure 3** Box A). The presence or primers will reduce the quality and yield of the sequencing run. If present, they can be removed by repeating the AMPure XP bead clean-up described in steps vii to xvii of the PCR amplification of bisulfite-converted library protocol. Make sure that molecular grade water is added to the library to adjust to a final volume of 50 μL before adding the 45 μL of AMPure beads. Once the library passes the QC, it can be stored until sequenced. For short-term storage, keep at −20°C, for longer-term keep at −80°C.

xix. Sequence the final library using the HiSeq Illumina platform.

xx. Analyse sequencing results.

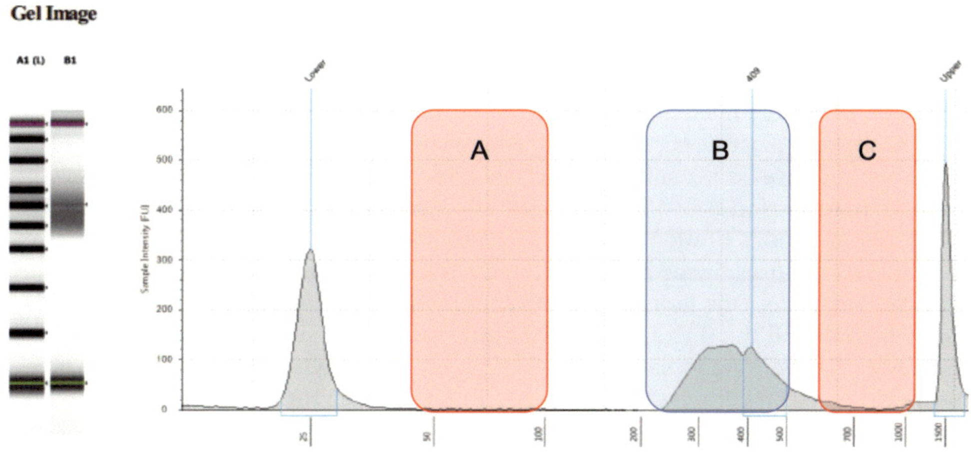

Figure 3.
Example electropherogram of successful WGBS library. Gel image on the left of the figure includes the gel images for (A1) the internal ladder and (B1) the WGBS library. The electropherogram on the right shows the lower and upper fragments of the internal ladder and the fragment size distribution for the WGBS (highlighted in blue in box B). The presence of peaks below 100 bp in the electropherogram is indicative of sequencing adapters or PCR primers. The presence of DNA fragments over 500 bp (Box C) indicates large fragments of DNA that could reduce the quality and output of the sequencing run. Both types of fragments should be removed using AMP XP beads size selection.

5. Data analysis and results

i. Perform FastQC Analysis to remove low-quality sequences.

ii. Use Trim Galore! (http://www.bioinformatics.babraham.ac.uk/projects/trim_galore) to trim sequencing adapters and to remove low-quality sequence.

iii. Perform FastQC Analysis to remove low-quality trimmed sequences.

iv. Map trimmed reads using Bismark aligner.

v. Remove PCR duplicates with Bismark Deduplicate function.

vi. Obtain methylation calls and methylation percentages per each CpG site using the Bismark Methylation Extractor function.

6. Conclusion

By following the protocol described herein, you have have a single-base resolution methylome for your sample. The quality of this methylome will depend on two main factors: (a) the sequencing depth of the produced methylome and (b) the number of replicates included in your experiment. With this data, you can infer methylation density at different genomic levels (i.e. along chromosomes; in different genomic features like genes, transposable elements, etc.) and within specific genomic features like promoters and gene bodies. If you are trying to identify changes in DNA methylation associated to a specific variable (e.g. growing environment, stress, tissue/cell type, age, disease, etc.), then you can identify differentially methylated cytosines (DMCs) or differentially methylated regions (DMRs) between groups of samples (i.e. control vs treatment). Methods such as Fisher's exact test can be used in the absence of replicates [34]. However, this approach does not consider the possibility of biological variability which is of great importance on a plastic trait such as DNA methylation. Linear or logistic regression-based methods are better suited to capture biological variability since they can compare methylation levels between groups of samples. One example of linear regression method is BSmooth [35] which assumes that data follows a binomial distribution and uses linear regression and t-tests to identify methylation differences for each site. One issue with linear regression is overfitting of DNA methylation levels beyond the 0 to 1 range that methylation proportion/fraction values regenerate. Logistic regression methods, implemented by software such as methylKit can deal better with data restricted to a 0 to 1 range by correcting to data dispersion.

Acknowledgements

This work is supported by the National Institute of Food and Agriculture, U.S. Department of Agriculture, Hatch Program number 2352987000.

Conflict of interest

The authors declare no conflict of interest.

Author details

Kendall R. Corbin and Carlos M. Rodriguez Lopez*
Environmental Epigenetics and Genetics Group, Department of Horticulture, College of Agriculture, Food and Environment, University of Kentucky, Lexington, KY, USA

*Address all correspondence to: cmro267@uky.edu

References

[1] Verhoeven KJF, Preite V. Epigenetic variation in asexually reproducing organisms. Evolution. 2014;**68**(3): 644-655

[2] Latzel V, Rendina González AP, Rosenthal J. Epigenetic memory as a basis for intelligent behavior in clonal plants. Frontiers in Plant Science. 2016; 7:1354

[3] Verhoeven KJF, Jansen JJ, van Dijk PJ, Biere A. Stress-induced DNA methylation changes and their heritability in asexual dandelions. New Phytologist. 2010;**185**(4):1108-1118

[4] Ocaña J, Walter B, Schellenbaum P. Stable MSAP markers for the distinction of *Vitis vinifera* cv Pinot noir clones. Molecular Biotechnology. 2013;**55**(3): 236-248

[5] Zhang Y-Y, Fischer M, Colot V, Bossdorf O. Epigenetic variation creates potential for evolution of plant phenotypic plasticity. New Phytologist. 2013;**197**(1):314-322

[6] Baránek M, Čechová J, Raddová J, Holleinová V, Ondrušíková E, Pidra M. Dynamics and reversibility of the DNA methylation landscape of grapevine plants (*Vitis vinifera*) stressed by in vitro cultivation and thermotherapy. PLoS One. 2015;**10**(5):e0126638

[7] Rodríguez López CM, Wilkinson MJ. Epi-fingerprinting and epi-interventions for improved crop production and food quality. Frontiers in Plant Science. 2015;**6**:397

[8] Vanyushin BF, Ashapkin VV. DNA methylation in higher plants: Past, present and future. Biochimica et Biophysica Acta. 2011;**1809**(8):360-368

[9] Herrera CM, Bazaga P. Untangling individual variation in natural populations: Ecological, genetic and epigenetic correlates of long-term inequality in herbivory. Molecular Ecology. 2011;**20**(8):1675-1688

[10] Tricker PJ, Gibbings JG, Rodríguez López CM, Hadley P, Wilkinson MJ. Low relative humidity triggers RNA-directed de novo DNA methylation and suppression of genes controlling stomatal development. Journal of Experimental Botany. 2012;**63**(10): 3799-3813

[11] Rico L, Ogaya R, Barbeta A, Peñuelas J. Changes in DNA methylation fingerprint of Quercus ilex trees in response to experimental field drought simulating projected climate change. Plant Biology. 2014;**16**(2): 419-427

[12] Ashikawa I. Surveying CpG methylation at 5′-CCGG in the genomes of rice cultivars. Plant Molecular Biology. 2001;**45**(1):31-39

[13] Humbeck K. Epigenetic and small RNA regulation of senescence. Plant Molecular Biology. 2013;**82**(6):529-537

[14] Tricker PJ, López CMR, Gibbings G, Hadley P, Wilkinson MJ. Transgenerational, dynamic methylation of stomata genes in response to low relative humidity. International Journal of Molecular Sciences. 2013;**14**(4):6674-6689

[15] Tricker P, Rodríguez López C, Hadley P, Wagstaff C, Wilkinson M. Pre-conditioning the epigenetic response to high vapor pressure deficit increases the drought tolerance of *Arabidopsis thaliana*. Plant Signaling & Behavior. 2013;**8**(10)

[16] Amoah S, Kurup S, Rodriguez Lopez CM, Welham SJ, Powers SJ, Hopkins CJ, et al. A hypomethylated population of *Brassica rapa* for forward

and reverse epi-genetics. BMC Plant Biology. 2012;**12**:193

[17] Róis AS, Rodríguez López CM, Cortinhas A, Erben M, Espírito-Santo D, Wilkinson MJ, et al. Epigenetic rather than genetic factors may explain phenotypic divergence between coastal populations of diploid and tetraploid Limonium spp. (Plumbaginaceae) in Portugal. BMC Plant Biology. 2013;**13**:205

[18] Rodríguez López CM, Wetten AC, Wilkinson MJ. Progressive erosion of genetic and epigenetic variation in callus-derived cocoa (*Theobroma cacao*) plants. New Phytologist. 2010;**186**(4):856-868

[19] Konate M, Wilkinson MJ, Mayne B, Scott E, Berger B, Lopez CMR. Atlas of age- and tissue-specific DNA methylation during early development of barley (*Hordeum vulgare*). Plant Sciences. 2018

[20] Konate M, Wilkinson MJ, Mayne BT, Pederson SM, Scott ES, Berger B, et al. Salt stress induces non-CG methylation in coding regions of barley seedlings (*Hordeum vulgare*). Plant Sciences. 2018

[21] Kitimu S, Taylor J, March T, Tairo F, Wilkinson M, Lopez R, et al. Meristem micropropagation of cassava (*Manihot esculenta*) evokes genome-wide changes in DNA methylation. Frontiers in Plant Science. 2015

[22] Boers R, Boers J, de Hoon B, Kockx C, Ozgur Z, Molijn A, et al. Genome-wide DNA methylation profiling using the methylation-dependent restriction enzyme LpnPI. Genome Research. 2018;**28**(1):88-99

[23] Rodríguez López CM, Guzmán Asenjo B, Lloyd AJ, Wilkinson MJ. Direct detection and quantification of methylation in nucleic acid sequences using high-resolution melting analysis. Analytical Chemistry. 2010;**82**(21): 9100-9108

[24] López CMR, Lloyd AJ, Leonard K, Wilkinson MJ. Differential effect of three base modifications on DNA thermostability revealed by high resolution melting. Analytical Chemistry. 2012;**84**(17):7336-7342

[25] Geyer KK, Rodríguez López CM, Chalmers IW, Munshi SE, Truscott M, Heald J, et al. Cytosine methylation regulates oviposition in the pathogenic blood fluke *Schistosoma mansoni*. Nature Communications. 2011;**2**:424

[26] Ellison A, Rodríguez López CM, Moran P, Breen J, Swain M, Megias M, et al. Epigenetic regulation of sex ratios may explain natural variation in self-fertilization rates. Proceedings. Biological sciences. 2015;**282**(1819)

[27] Guerrero-Bosagna C. DNA methylation research methods. Materials and Methods. 2013;**1**:3

[28] Plongthongkum N, Diep DH, Zhang K. Advances in the profiling of DNA modifications: Cytosine methylation and beyond. Nature Reviews Genetics. 2014;**15**:647-661

[29] Kurdyukov S, Bullock M. DNA methylation analysis: Choosing the right method. Biology (Basel). 2016;**5**(1)

[30] Coombe BG. Growth stages of the grapevine: Adoption of a system for identifying grapevine growth stages. Australian Journal of Grape and Wine Research. 1995;**1**(2):104-110

[31] Sahu SK, Thangaraj M, Kathiresan K. DNA extraction protocol for plants with high levels of secondary metabolites and polysaccharides without using liquid nitrogen and phenol. ISRN Molecular Biology. 2012; **2012**:205049

[32] Gehring M, Henikoff S. DNA methylation dynamics in plant genomes. Biochimica et Biophysica Acta. 2007; **1769**(5–6):276-286

[33] Lodhi MA, Ye GN, Weeden NF, Reisch B. A simple and efficient method for DNA extraction from grapevine cultivars and *Vitis* species. Plant Molecular Biology Reporter. 1994;**12**(1): 6-13

[34] Wreczycka K, Gosdchan A, Yusuf D, Grüning B, Assenov Y, Akalin A. Strategies for analyzing bisulfite sequencing data. Journal of Biotechnology. 2017;**261**:105-115

[35] Hansen KD, Langmead B, Irizarry RA. BSmooth: From whole genome bisulfite sequencing reads to differentially methylated regions. Genome Biology. 2012;**13**(10):R83